Lecture Notes in Networks and Systems

Volume 453

The series "Lecture Notes in Networks and Systems" publishes the latest developments in Networks and Systems—quickly, informally and with high quality. Original research reported in proceedings and post-proceedings represents the core of LNNS.

Volumes published in LNNS embrace all aspects and subfields of, as well as new challenges in, Networks and Systems.

The series contains proceedings and edited volumes in systems and networks, spanning the areas of Cyber-Physical Systems, Autonomous Systems, Sensor Networks, Control Systems, Energy Systems, Automotive Systems, Biological Systems, Vehicular Networking and Connected Vehicles, Aerospace Systems, Automation, Manufacturing, Smart Grids, Nonlinear Systems, Power Systems, Robotics, Social Systems, Economic Systems and other. Of particular value to both the contributors and the readership are the short publication timeframe and the world-wide distribution and exposure which enable both a wide and rapid dissemination of research output.

The series covers the theory, applications, and perspectives on the state of the art and future developments relevant to systems and networks, decision making, control, complex processes and related areas, as embedded in the fields of interdisciplinary and applied sciences, engineering, computer science, physics, economics, social, and life sciences, as well as the paradigms and methodologies behind them.

Indexed by SCOPUS, INSPEC, WTI Frankfurt eG, zbMATH, SCImago.

All books published in the series are submitted for consideration in Web of Science.

For proposals from Asia please contact Aninda Bose (aninda.bose@springer.com).

More information about this series at https://link.springer.com/bookseries/15179

Phayung Meesad · Sunantha Sodsee ·
Watchareewan Jitsakul ·
Sakchai Tangwannawit
Editors

Proceedings of the 18th International Conference on Computing and Information Technology (IC2IT 2022)

Springer

Editors
Phayung Meesad
Faculty of IT and Digital Innovation
King Mongkut's University of Technology North Bangkok
Bangkok, Thailand

Sunantha Sodsee
Faculty of IT and Digital Innovation
King Mongkut's University of Technology North Bangkok
Bangkok, Thailand

Watchareewan Jitsakul
Faculty of IT and Digital Innovation
King Mongkut's University of Technology North Bangkok
Bangkok, Thailand

Sakchai Tangwannawit
Faculty of IT and Digital Innovation
King Mongkut's University of Technology North Bangkok
Bangkok, Thailand

ISSN 2367-3370 ISSN 2367-3389 (electronic)
Lecture Notes in Networks and Systems
ISBN 978-3-030-99947-6 ISBN 978-3-030-99948-3 (eBook)
https://doi.org/10.1007/978-3-030-99948-3

This Springer imprint is published by the registered company Springer Nature Switzerland AG
The registered company address is: Gewerbestrasse 11, 6330 Cham, Switzerland

Preface

The COVID-19 pandemic around the world has greatly affected our everyday lives. Human has to adapt its way of living to prevent the spread of the COVID-19. There is no denying that digital technology has become an essential tool that has facilitated our lives during this hard time. Advances in computer networks and information systems have inspired new innovations around the world that will help us through this crisis and empower humankind in this turning point of history.

As a sharing platform, the International Conference on Computing and Information Technology (IC2IT) has been organized for eighteen years to serve digital communities in exchanging ideas, knowledge, and researches to inspire new digital innovations. In 2022, the IC2IT is held on May 19–20 in Kanchanaburi, Thailand. The main contributions in the conference proceedings are in the areas of machine learning, natural language processing, image processing, intelligent systems and algorithms, as well as network and cloud computing. The paper submissions were from nine countries and were peer reviewed by at least three members of the program committee. Among these submissions, eighteen have been accepted for publication based on originality and scientific result. All accepted papers are included in the conference proceedings, which are published in the well-established and worldwide-distributed Lecture Notes in Networks and Systems series edited by Janusz Kacprzyk.

Herein, the organizing committee would like to thank the authors, program committee, and all collaborating institutions in both Thailand and overseas for their outstanding support and academic cooperation, as well as the staff members of the Faculty of Information Technology and Digital Innovation at King Mongkut's University of Technology North Bangkok.

February 2022

Phayung Meesad
Sunantha Sodsee
Watchareewan Jitsakul
Sakchai Tangwannawit

Organization

Program Committee

S. Auwatanamongkol	NIDA, Thailand
T. Bernard	Li-Parad, France
S. Boonkrong	SUT, Thailand
P. Boonyopakorn	KMUTNB, Thailand
V. Chongsuphajaisiddhi	KMUTT, Thailand
N. H. H. Cuong	UD, Vietnam
N. Ditcharoen	UBU, Thailand
T. Eggendorfer	HS, Weingarten, Germany
K. Hengpraprohm	NPRU, Thailand
S. Hengpraprohm	NPRU, Thailand
S. Hiranpongsin	UBU, Thailand
T. Jensuttiwetchakul	KMUTNB, Thailand
W. Jitsakul	KMUTNB, Thailand
N. Jongsawat	RMUTT, Thailand
M. Kaenampornpan	MSU, Thailand
M. Ketcham	KMUTNB, Thailand
P. Kropf	Uni NE, Switzerland
M. Kubek	FernUni, Germany
P. Kunakornvong	RMUTT, Thailand
K. Laosen	PSU, Thailand
U. Lechner	UniBW, Germany
M. Maliyaem	KMUTNB, Thailand
P. Meesad	KMUTNB, Thailand
S. Nuchitprasitchai	KMUTNB, Thailand
K. Pasupa	KMITL, Thailand
S. Pattanavichai	RMUTT, Thailand
S. Plaengsorn	PBRU, Thailand
J. Polpinij	MSU, Thailand

N. Porrawatpreyakorn	KMUTNB, Thailand
P. Saengsiri	TISTR, Thailand
T. Sarawong	RMUTK, Thailand
T. Siriborvornratanakul	NIDA, Thailand
S. Smanchat	KMUTNB, Thailand
S. Sodsee	KMUTNB, Thailand
G. Somprasertsri	MSU, Thailand
W. Sriurai	UBU, Thailand
T. Sucontphunt	NIDA, Thailand
A. Taguchi	TCU, Japan
S. Tangwannawit	KMUTNB, Thailand
J. Thongkam	MSU, Thailand
N. Tongtep	PSU, Thailand
D. Tutsch	Uni-Wuppertal, Germany
N. Utakrit	KMUTNB, Thailand
N. Utakrit	KMUTNB, Thailand
S. Valuvanathorn	UBU, Thailand
K. Viriyapant	KMUTNB, Thailand
P. Wuttidittachotti	KMUTNB, Thailand

Organizing Partners

In Cooperation with
King Mongkut's University of Technology North Bangkok (KMUTNB)
FernUniversitaet in Hagen, Germany (FernUni)
Chemnitz University of Technology, Germany (CUT)
Oklahoma State University, USA (OSU)
Edith Cowan University, Western Australia (ECU)
Hanoi National University of Education, Vietnam (HNUE)
Mahasarakham University (MSU)
Nakhon Pathom Rajabhat University (NPRU)
National Institute of Development Administration (NIDA)
Rajamangala University of Technology Thanyaburi (RMUTT)
Rajamangala University of Technology Krungthep (RMUTK)
Kanchanaburi Rajabhat University (KRU)
Phetchaburi Rajabhat University (PBRU)
Gesellschaft für Informatik (GI)
Council of IT Deans of Thailand (CITT)
The Thailand Convention and Exhibition Bureau (TCEB)

Contents

Automatical IVUS Image Segmentation and Detection Using BLOB Analysis

Nippitch Onhirun and Pattama Charoenporn(✉)

Department of Computer Science, King Mongkut's Institute of Technology Ladkrabang, Bangkok, Thailand
{61605040,Pattama.ch}@kmitl.ac.th

Abstract. In this research, experiments were conducted on the problem of image analysis in atherosclerotic disease. The study was conducted from intravascular ultrasound images that have been taken inside the blood vessels from patients with arterial wall hardening or stenosis. These images are used by experts or specialists to analyze the problem of the disease. There may be some interference from shadows or elements that occurred while taking the image that may cause the incomplete image. We propose an applied method by using computer techniques that help to analyze the components within the image to find the area of interest by enhancement image. Then, the pixel levels are analyzed and classified to divide the background from the image. Next, the morphological operation is used to readjust image properties, and blob analysis is used to identify the region of interesting pixel values. In the final process, Connected Component Analysis (CCA) was used to detect media-adventitia and lumen boundary area The efficacy of segment results was measured by comparing with the expert to measure performance as (media-adventitia, lumen) the Jaccard Index = (0.9570, 0.96695), Hausdorff Distance = (0.5259, 0.6304), Percentage Area Distance = (0.0382, 0.03395) IVUS dataset was used from Simone Balocco, Dept. Matemàtica Aplicada i Anàlisi, Universitat de Barcelona, Barcelona, Spain.

Keywords: Blob analysis · Segmentation · Atherosclerotic disease · IVUS image

1 Introduction

Atherosclerosis is a vascular disease [1] characterized by the thickening of the walls of blood vessels. Plaque is formed by an accumulation of fat and fibrous tissues, and this leads to the narrowing of the blood vessels. This plaque may lead to less blood supply to the peripheral organs and eventually necrosis of them. Atherosclerosis is a chronic disease that is common in the elderly, and it is the leading cause of death in countries around the world after cancer and accidents. Heart disease is caused by the presence of plaque clinging to the walls of blood vessels, blocking the flow of blood, especially when clogged coronary arteries that supply the blood flow to a heart are reduced. When examining the plaque, that it is full of cholesterol fats mixed with cells, cell debris, fibrous proteins, and calcium. The plaque may gradually accumulate and increase in

P. Meesad et al. (Eds.): IC2IT 2022, LNNS 453, pp. 1–10, 2022.
https://doi.org/10.1007/978-3-030-99948-3_1

size, causing the artery to narrow, blocking the flow of blood, or it may rupture until it clogs the blood vessels at that time. This hardening leads to narrowing of the blood vessels resulting in less blood supply to the peripheral organs and eventually the death of the peripheral organs. This causes the need to cut off the said organ or may cause the patient's death. If an expert can early diagnose and treat this morbidness, it will reduce the mortality rate.

The method for distinguishing atherosclerosis is an image-processing study that identifies different regions of the visual element, e.g., lumen, media, adventitia. These regions on the artery wall are regions of interest because they are where the plaque was built up or they can be used to indicate the abnormalities. This research focuses on separate things that were mentioned above apart to identify the areas that make the diagnosis more convenient. However, the previous study found that there are interferences for image analysis on IVUS which have been caused by several factors such as the shadows that occur during taking images of the arterial junction at the imaging site. These reduce the accuracy of the results.

In the previous research for detecting the calcification boundary area, Hanah Sofian et al. [2] proposed the Otsu threshold, morphological operation and empirical threshold value to segmented media-adventitia boundary. Janya O. et al. [3] used ellipse analysis to estimate areas and apply the Principal Component Analysis (PCA) method to locate the media adventitia regions. Anusorn W. et al. [4] proposed method to segment Media-Adventitia and lumen boundary. IVUS image was used to adjust the intensity with contrast stretching, then classifying intensity with adaptive clustering and preprocessing intensity into a binary image. Then, the image was segmented the area and detected contour by convex hull.

2 Background Knowledge

2.1 Image Description

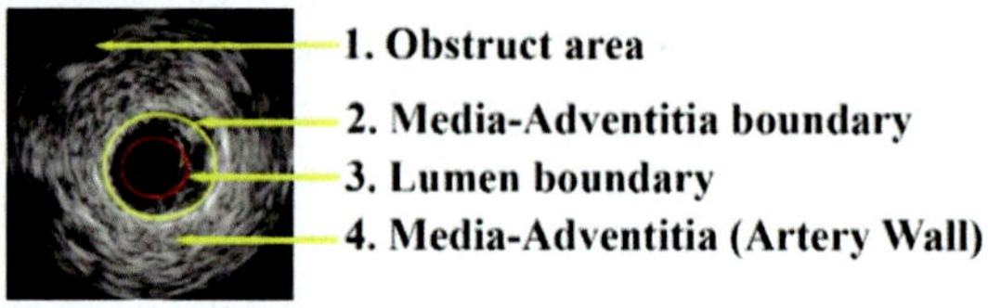

Fig. 1. IVUS image.

According to Fig. 1. Plaque commonly forms in the intima layer where located between the lumen boundary and media adventitia boundary. This plaque caused it more difficult to identify layers of blood vessels. This figure is divided into four areas: The image area where shadows or noise results in an incomplete image, the outer vascular wall. Part of the inner vascular wall, part of the artery wall.

The region of interest is the area between the boundary of the artery wall and the lumen of the artery. In this research, the focus is on the identification of the blood flowing area of the image with imperfections from shadows and other elements.

2.2 Enhancement Methodology

Image enhancement [5] is a process of improving values pixel in the grayscale image to perform the data quality. It is one of the steps to optimize a data value in the image via improving the quality of the image to a better level of detail and clearly before being analyzed. These result in more quality images. The grayscale image was improved by using the histogram analysis process. The frequency of the image data in the histogram image represents the density characteristics of intensity values due to the darkness within the image. It is an adjustment in the sharpness of the image data by image stretching from the original intensity value in a grayscale range is separated until the maximum gray level that the computer can display (256 levels), resulting in more sharp images. Image stretching from the original intensity value in a grayscale range. This method expands the frequency range of the data within the histogram using linear mapping, which takes the highest mean from the histogram and adjusts the range of dense data in the dark to spread out. Linear contrast mapping is a simple method to adjust the grayscale level or the brightness value. By expanding the range of the gray level of the original data to be more valuable until the full range 0–255 by using a linear lookup table such as techniques standard deviation linear contrast stretch, minimum-maximum contrast stretch or data scaling.

2.3 Clustering Method

The clustering method [6] is a data grouping by divide data from their similarity. In addition, this method is used to classification data in groups depending on their intensity. The data with an intensity value of 0, 1–30, 31–100, more than 100 will be consecutively classified as groups 1, 2, 3, and 4. After clustering data, data that belong to groups 2, 3, and 4 will be used to set a threshold for converting images into binary images to identify the lumen area. Then data that belong to groups 3 and 4 will be used to set a threshold for converting another binary image to identify media-adventitia area (Fig. 2).

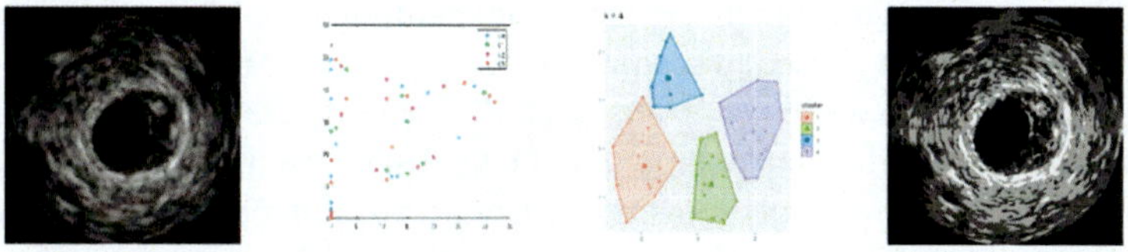

(a) (b) (c) (d)

Fig. 2. Classification process. (a) original image (b) scatter diagram that demonstrate the distribution intensity value (c) scatter diagram that demonstrate value of each group (d) result of clustered image divide intensity to 4 groups.

2.4 Morphological Method

The morphological method is the process of integrate the processes of non-linear ordinary differential equations which are related to the contour of the characteristics structure in the image. Morphological operations are an image processing tool for erode or enhance the edges or structure of an image. By using the set theory, the set in morphology represents the shape of the object in the image such as a cluster of black and white within the binary image be used to remove noise and obstruct, dilate the area of the object to be larger.

3 Methodology

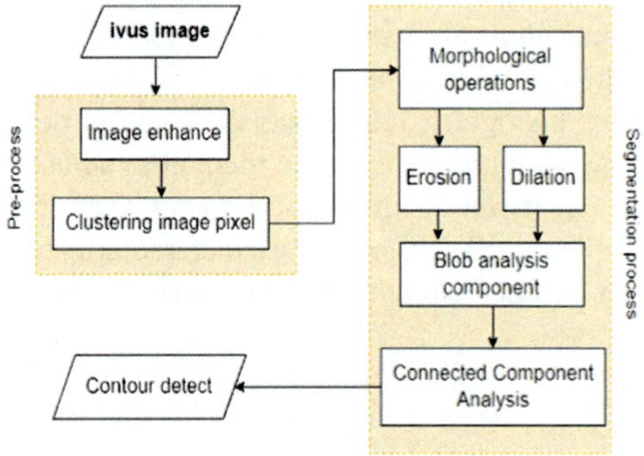

Fig. 3. Method diagram.

Figure 3 shows a diagram of all the processes to segment the IVUS image. In the Pre-process step, consist of are two processes are used. The first process is an image enhance to stretch the light within the image using an image stretching. Next process, the image was clustered pixel groups to define the component in the image and classify the pixel in 4 groups from the image. Then, converting the image data obtained from the cluster into a binary image before the Segmentation step. The Segmentation step is segmenting the area of interest that is the lumen and media-adventitia. There are three processes, the Morphological operation is used to readjust image properties. After that, Blob analysis [7] used to segment the image pixel that has already been extracted from the background. By the black pixel was identified position. In the final step, the image was segmentation with the most connected white pixels using the Connected Component Analysis method. The contour is obtained. Describing in Fig. 6.

There are 2 processes in this method; Pre-process step and Segmentation process. Pre-process step, to improve the image properties to adjust the component within the image. In the preprocess step, the image is extracted into two parts: lumen and media adventitia, dilation method was used to identify lumen. Erosion method was used to identify media-adventitia. Segmentation process, we bring the image obtained by adjusting the component within the image from Pre-process step after classified the group of pixel value convert to binary image.

3.1 Morphological Operations

- Dilation operation
  The Dilation method [8] describing the pixel value from the structuring element each value of the pixel is defined by checking the pixel of the position on the image. After checking the pixel, the pixel value of 0 is changed to 1 in the area around the object.

$$D(x, y) = Dilate\, f(x, y) \oplus ST_1 \tag{1}$$

  $D(x, y)$ is a result of dilation process and ST_1 is a disk of the structure element.
- Erosion operation
  On the other hand, the Erosion method is the opposite method of Dilate operation. Erosion method was used to identify media-adventitia area, by dilating image and then erode after. Equation as follows;

$$E(x, y) = Erode\, f(x, y) \ominus ST_2 \tag{2}$$

  $E(x, y)$ is a result of dilation process ST_1 is a square of the structure element.

3.2 Blob Detection

After image enhancement and clustering grayscale image. The next step, taking the image obtained from the classification process by converting the result of the image into a binary image and getting the image into the analysis of the components within the image. Media-adventitia and lumen boundaries are obtained from this process. Therefore, the region be specified for the segmentation process.

Figure 4. It can be seen that in the black area in the middle of the image, the pixel value 0 when using blob analysis to find the black pixel position. Continue to find black pixel from left to the right, top to bottom. When it finds a black pixel, these around the connect area until the whole image is completed process.

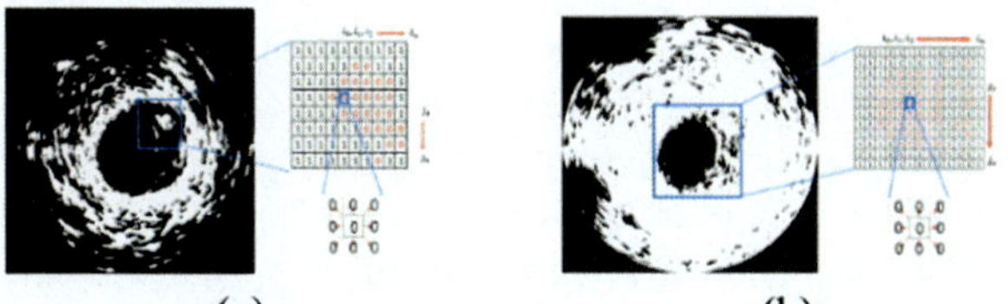

(a) (b)

Fig. 4. (a) Blob detecting of media-adventitia. (b) Blob detecting of lumen.

$$f_{(x,y)} = \begin{cases} [x_i, y_i] \Rightarrow x_i, y_i > 0, = 1 \\ 0 \qquad otherwise \end{cases} \tag{3}$$

$$\mu_i^x = \frac{1}{N}\left[\sum\nolimits_{i=1}^{n} x_i\right], \mu_j^y = \frac{1}{N}\left[\sum\nolimits_{j=1}^{n} y_j\right] \tag{4}$$

Where N is the number of pixels in the blob and x_i, y_j are the x and y coordinates of the N pixels, respectively. In situations where the blob contains. And scan on first to N pixel. Find the area and perimeter of the following shapes a blob is the length of the contour in the blob. This can be found by scanning along the boundary coronary artery (region of interest) and the whole of image an object and result in the number of pixels encountered.

```
Input
  IVUS binary after classify from cluster pixel f(x,y)
Compute
  To find all white pixel for scope area to analysis with blob
  To find pixel values all over image Σ = [xi,yj]
Looping
  Row    xi = 0 to width of image μi^x = 1/N [Σ_{i=1}^n xi ]
  Column yj = 0 to height of image μj^y = 1/N [Σ_{j=1}^n yj ]
  If (pixel == 1) then 1 label at xi,yj to value = 0
  Else return intensity = 0 then label at xi,yj to value = 1
Return
   Result pixel values from segment with blob f(x,y)
```

According to program code. The first step obtains the values from binary $f_{(x,y)}$, then finds all white pixels for all the area scope to analyze with blob. Next, to determine pixel value all over the image by looping x_i row equal to zero until a width of an image, looping y_j column is equal to zero and until the height of an image. If a pixel value is equal to 1 and to do position identify at x_i, y_j value equal to zero and then label this position. But the pixel is zero to do the value is 1. Lastly, obtain the result pixel values from the segment with Blob$f_{(x,y)}$.

3.3 Connected Component Analysis

In the next step, after extracting the image using the blob analysis method, these result are obtained image all pixel values of interest in the binary image. However, the result from the before method can identify component areas of interest. And this process was given result is incomplete. Therefore, the technique of Connected Component Analysis [9] has been applied to find connected regions which are most interconnected. The result in the acquisition of media-adventitia and lumen show from Fig. 5.

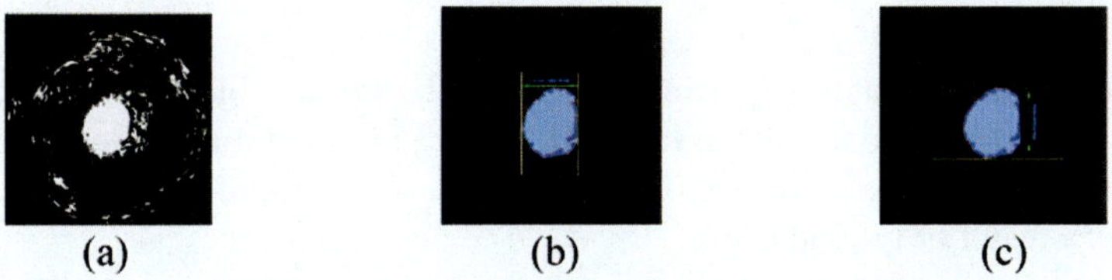

Fig. 5. The results segmented with blob and CCA. (a) The image after the blob analysis process. In the figure (b) and (c) an image with the line that calculates the maximum width and height that emerges around the object.

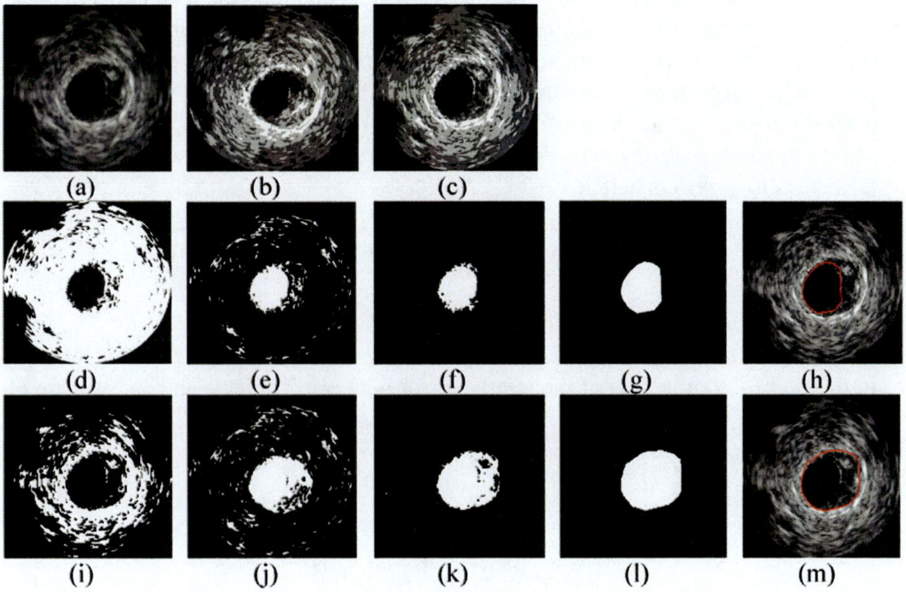

Fig. 6. (a) original image, (b) enhancement image, (c) clustered image, (d) result of binary image (value of clustered groups 2,3,4), (e, j) lumen and media-adventitia area obtained by Blob analysis, (f, g, k, l) lumen and media-adventitia area obtained by CCA (convex hull), (i) result of Binary image process by cluster value (value of clustered groups cluster group 3,4), (h, m) result of lumen and media-adventitia boundary from segment.

4 Results and Discussion

4.1 Performance Measurement Tools

In the performance measurement process the results of the image segmentation were obtained. Compare the boundaries between the results obtained from the proposed method and the results of the expert who segment the area on IVUS image. In this comparison, there were methods: 1. Jaccard Index 2. Hausdorff. Distance 3. Percentage Area Distance.

1. *Jaccard Index*
Jaccard Index (JI) is a principal verification similar for the object area that compares the result of the proposed method with the result of ground truth. Selected the maximum difference value is close to one, which M_{seg} is a result of the proposed method, M_{man} is a result of ground truth.

$$JI\left(M_{seg}, M_{man}\right) = 1 - \frac{|M_{seg} \cap M_{man}|}{|M_{seg} \cup M_{man}|} \tag{5}$$

2. *Hausdorff Distance*
Comparison with the Hausdorff Distance (HD) has obtained the result area to measurement similar with the result of ground truth. selected a minimum distance value has a between 2 results. Which M_{seg} is a result of the proposed method, M_{man} is a result of ground-truth, d is a sum of the scale with a comparison of 2 objects. Hausdorff distance is a measure performance result of segmentation by propose method were compared with manual segmentation of ground truth the result of Hausdorff Distance close zero is better.

$$HD\left(M_{seg}, M_{man}\right) = \left\{ \max_{a \in M_{seg}} [d(a, b)], \max_{b \in M_{man}} [d(a, b)] \right\} \tag{6}$$

3. *Percentage Area Distance*
Percentage Area Distance (PAD) is a measurement of the relative error object area. In this step, a comparison of the relative error between 2 objects and an estimate of the minimum difference value is close to zero. At a M_{seg} is a result of the proposed method, M_{man} is a result of the ground truth.

$$PAD = \left| \frac{M_{seg} - M_{man}}{M_{man}} \right| x100 \tag{7}$$

In this research experiment, the image algorithm was used to adjust the properties to obtain clear internal elements. Then, the result was taken to group with segmentation process. Next, the result from the segmentation process was pre-processed again with the morphological operation. Then, analyzing the properties of the area of interest, such as lumen and media-adventitia were performed by Blob analysis to extract both areas. The properties of acquired images were adjusted by CCA analysis, and the results were compared with the previous research, as shown in Table 1. The results of the fractionation efficiency by taking all the results obtained to find the standard deviation were determined as shown in Table 2.

However, from the experimental results with 24 images, in the proposed method got results value as compared with manual segment boundary by the expert from lumen boundary were obtain 0.9570 on Jaccard Index (JD) (Eq. (5)), 0.5291 on Hausdorff Distance (HD) Eq. (6) and 0.0382 at Percentage Area Distance (PAD) Eq. (7) JI when compared with the results, the method of this research approach is closer to 1, which is more accurate than the other methods, whereas HD approaches 0, therefore more accurate than the other methods. IVUS dataset [10] was used from Simone Balocco.

Table 1. Show the comparison between propose method and other.

Method	JI	HD	PAD
Anusorn w.	0.9453	0.5683	0.0167
H. Sofian	0.8541	0.7921	0.4360
Propose	0.9570	0.5291	0.0382

Table 2. STD (standard deviation), comparison media-adventitia and lumen boundary with results by expert.

IVUS IMAGE	Method (media-adventitia, lumen)		
	JI	HD	PAD
frame_02_0001_001	0.9667,0.9652	6.4772, 5.1524	0.0325, 0.0352
frame_02_0001_002	0.9542, 0.9637	5.7456, 5.2267	0.0371, 0.0335
frame_02_0001_003	0.9781, 0.9493	6.7958, 4.8276	0.0311, 0.0304
frame_02_0001_004	0.9672, 0.9621	5.8827, 5.3254	0.0337, 0.0443
frame_02_0001_005	0.9833, 0.9587	6.2426, 4.8998	0.0212, 0.0371
frame_02_0002_004	0.9591, 0.9565	5.9878, 5.4225	0.0324, 0.0359
frame_02_0002_005	0.9503, 0.8797	5.9667, 3.8899	0.0811, 0.0486
frame_02_0003_001	0.9643, 0.8848	5.2915, 4.8733	0.0233, 0.0592
frame_02_0003_002	0.9795, 0.9392	6.6744, 5.6293	0.0132, 0.0454
frame_04_0048_003	0.9698, 0.9564	5.7689, 5.8827	0.0457, 0.0415
frame_04_0048_005	0.9424, 0.9673	5.4882, 4.9884	0.0342, 0.0537
frame_04_0050_003	0.9711, 0.9681	6.3654, 4.5953	0.0116, 0.0546
frame_04_0050_004	0.9472, 0.9437	5.6569, 4.6585	0.0332, 0.0622
frame_04_0050_005	0.9658, 0.9486	6.3838, 4.9983	0.0431, 0.0524
frame_07_0018_003	0.9759, 0.9563	6.4417, 5.5896	0.0367, 0.0467
frame_07_0018_004	0.9633, 0.9739	5.9604, 5.7965	0.0382, 0.0374
frame_07_0018_005	0.9712, 0.9576	6.5872, 5.8447	0.0384, 0.0226
frame_07_0019_001	0.9592, 0.9476	5.9878, 5.4981	0.0471, 0.0355
frame_07_0019_002	0.9753, 0.9598	6.4332, 5.8843	0.0362, 0.0319
frame_07_0019_003	0.9765, 0.9587	6.4793, 4.8748	0.0367, 0.0275
frame_07_0019_004	0.9597, 0.9351	5.9057, 5.9962	0.0244, 0.0412
frame_07_0019_005	0.9724, 0.9586	6.4238, 5.8957	0.0133, 0.0368
frame_07_0020_005	0.9686, 0.9694	6.5543, 4.7574	0.0121, 0.0391
frame_07_0021_003	0.9644, 0.9503	6.4472, 5.2916	0.0412, 0.0322
STD of media boundary	0.010388531	0.39975791	0.014656378
STD of lumen boundary	0.023046964	0.52674492	0.010143827

5 Conclusion

In the conclusion, this research is applied methods are used to segmentation region of interest in IVUS images. Consist of the following methods are image enhancement to adjust properties within the image, image cluster pixels to classify the data of grayscale image, the image properties were then adjusted using morphology. Then, blob analysis to segment the lumen and media-adventitia boundary from the image. In the final step, the extracted images are taken and analysis to detecting contour with the CCA. That may contribute to supporting the expert or physician to decision in the diagnosis. In the future, we will improve the segmentation methods and results to be more efficient and more accurate.

References

1. Elangovan, V.R., Rajeswari, A.J., Akila, D., Hema Shankari, K., Suseendran, G.: Heart atherosclerosis detection using FCM+kMeans Algorithm. In: 2nd International Conference on Computation, Automation and Knowledge Management (ICCAKM), IEEE, Dubai (2021)
2. Sofian, H., Ng, A., Than, J., Mohamad, S., Noor, N.M.: Calcification boundary detection in coronary artery using intravascular ultrasound images. In: The 2017 IEEE Region 10 Conference (TENCON), pp. 2835–2839. IEEE, Penang (2017)
3. Onpans, J., Yookwan, W., Srikamdee, S.: Automated media segmentation in intravascular ultrasound images using geometric principal component analysis. In: 18th International Conference on Electrical Engineering/Electronics, Computer, Telecommunications and Information Technology, pp. 853–856. IEEE, Chiang Mai (2021)
4. Wong-Od, A., Rodtook, A., Rasmequan, S., Chinnasarn, K.: Automated segmentation of media-adventitia and lumen from intravascular ultrasound images using non-parametric thresholding. In: 9th International Conference on Knowledge and Smart Technology, pp. 220–225. IEEE, Chonburi (2017)
5. Yelmanov, S., Romanyshyn, Y.: Image enhancement technique for use in real-time mobile applications. In: 39th International Conference on Electronics and Nanotechnology (ELNANO), pp. 553–558. IEEE, Kyiv (2019)
6. Wang, J., Su, X.: An improved K-Means clustering algorithm. In: 3rd International Conference on Communication Software and Networks, pp.44–46. IEEE, Xi'an (2011)
7. Bedruz, R.A., Sybingco, E., Bandala, A., Quiros, A.R., Uy, A.C., Dadios, E.: Real-time vehicle detection and tracking using a mean-shift based blob analysis and tracking approach. In: 9th International Conference on Humanoid, Nanotechnology, Information Technology, Communication and Control, Environment and Management (HNICEM), pp. 1–5. IEEE, Manila (2017)
8. De Natale Francesco, G.B., Boato, G.: Detecting morphological filtering of binary images. IEEE Trans. Inf. Forens. Secur. **12**(5), 1207–1217 (2017)
9. Noknaka, K., Sankoh, H., Wanabe, R., Sabirin, H., Naito, S.: Efficient parallel connected component labeling with a coarse-to-fine strategy. IEEE Access **6**, 55731–55740 (2018)
10. Balocco, S., et al.: Standardized evaluation methodology and reference database for evaluating IVUS image segmentation. Comput. Med. Imaging Graph. **38**(2), 70–90 (2013)

Multi-tenant Blockchain-Based State Land Allocation and Usage System for Government Agencies

Wee Sritippho and Sukumal Kitisin(✉)

Department of Computer Science, Faculty of Science, Kasetsart University, Bangkok, Thailand
{wee.s,sukumal.i}@ku.th

Abstract. In many countries, vacant or underutilized state land would be allocated by the government to their citizens for specific usages. Digital records on these allocations might be managed by multiple agencies in different ministries, which may not be accessible across agencies. This could potentially cause data inconsistency and have a negative impact on the overall data quality. Blockchain, the technology that allows multiple entities to reliably manage a single dataset in a distributed manner together, coupled with Geographic Information System, can improve data consistency of the allocated state lands and their usages. A prototype of blockchain-based land allocation system is proposed to facilitate collaboration in state land allocation data management among multiple agencies.

Keywords: Blockchain · Ethereum · Geographic Information System · Government · Smart contract · State land allocation

1 Introduction

State land allocation is a process that governments allocate vacant or underutilized state land to use by citizens for specific usages—e.g., residential, industrial, and agriculture. Different countries have different laws and policies about state land allocation. Some countries might have a single agency that supervises every state land, whereas other countries might have multiple agencies that supervise multiple types of state land. For example, Thailand has Royal Forest Department that manages state lands in reserved forest boundary, Agricultural Land Reform Office that manages land reformed areas, Treasury Department that manages royal properties, and so forth.

For countries that have multiple agencies to supervise state lands, maintaining data consistency among multiple agencies is a challenging task—e.g., an agency might allocate an area of land that geographically overlaps with another area of land that was already allocated by another agency. Ensuring state land data consistency among multiple agencies would improve the overall data quality, thereby providing better input for national decision-making. However, due

P. Meesad et al. (Eds.): IC2IT 2022, LNNS 453, pp. 11–21, 2022.
https://doi.org/10.1007/978-3-030-99948-3_2

to unexpected errors during some processes (e.g., cadastral survey, communication, or data conversion), state land allocation data from one agency might geographically overlap with others, leading to conflict and distrustfulness between agencies. One solution is to formally establish an agency—which is trusted by every agency—to centralize the state land allocation data. Another solution is to implement a distributed trust mechanism which can cross-check the land allocation state across agencies.

This research aims at creating a blockchain-based state land allocation system that could act as a single source of truth for multiple government agencies that don't fully trust each other. The system mainly uses a smart contract that operates in a public permissioned blockchainnetwork—an isolated network that only comprises nodes from designated participants. Relational databases are also used to improve overall performance and fulfill the requirements that can't be accomplished by using blockchain alone.

2 Related Works and Technologies

2.1 Blockchain

Blockchain is a combination of technologies that provide a shared immutable ledger that can be used by different parties in a distributed manner. Its most notable usage is the Bitcoin cryptocurrency, which can be used to trade for goods and services without relying on trusted intermediaries [1], i.e., banks.

A Block in blockchain is basically a set of transactions bundled together and supplemented with other information, e.g., timestamp, its hash, and previous block's hash. Usually, information to prevent block tampering, e.g., nonce (for proof-of-work blockchains) or the signer's signature (for proof-of-authority), would also be added to the block. By including the previous block's hash, blocks would logically be connected to each other like a chain, hence the name "blockchain".

Blockchains can be categorized as public permissionless blockchains, public permissioned blockchains, and private blockchains [2].

- Public permissionless blockchains are open systems where all users may join and leave the network at will. The best example for this category is Bitcoin, where anyone can generate a wallet and join the network, or even use their computer to act as a node in the network.
- Public permissioned blockchains were introduced to control which users can join the network along with their write and read access rights. Examples of public permissioned blockchains are Hyperledger Fabric [3] and Corda [4].
- Private blockchains are usually controlled by a single organization. They are comparable to traditional centralized systems running cryptographic protocols. Examples are Multichain and Eris.

2.2 Ethereum

Ethereum [5] is an open-source blockchain platform that allows participants to create and use smart contracts—computer programs that can be deployed and executed on an Ethereum blockchain network. Ethereum smart contracts can be written in smart contract languages, e.g., Solidity, Vyper, Yul, and Fe, which are similar to programming languages. An application can be built on top of smart contracts. Such an application would be called decentralized application (dApp).

Ethereum platform focuses on the primary network called Mainnet. In addition to the primary network, there are also other independent networks that are used for testing and smart contract development purpose, e.g., Görli, Kovan, Rinkeby, and Ropsten. Apart from using public networks. It is also possible to create a private network by using Ethereum client software.

2.3 Geographic Information System (GIS)

Geographic Information System (GIS) [6] is a computer system that specializes in capturing, storing, verifying, and displaying geospatial data—data that include location information, usually places across the world. Locations on earth can be described in many different ways, e.g., latitude and longitude, address, or ZIP code.

GIS data formats can be classified into two types, i.e., raster and vector. Raster format is useful for storing information that can be represented by a matrix of cells, e.g., elevation and satellite images. Vector format can store vertices and paths, which are suitable for storing data that represent precise coordinates or boundaries. Three fundamental elements of the vector format are points, lines and polygons (areas).

2.4 PostGIS

PostGIS [7] is an open-source extension for the PostgreSQL [8] open-source relational database management system (RDBMS) which adds many geographic database types, functions, operators, and index enhancements, transforming PostgreSQL into a geospatial RDBMS. It is supported in most operating systems, e.g., Linux, Unix, Windows, macOS, and any platforms supported by PostgreSQL.

2.5 Blockchain-Based GIS

Blockchain-based GIS has been studied in many contexts, including land registration, border violation detection, and food tracking systems (Mendi et al., 2020) [9].

Farnaghi and Mansourian (2020) [10] introduced a decentralized public participatory GIS (PPGIS) which encodes the spatial data as GeoJSON and then saves the GeoJSON as a string on an Ethereum smart contract. They also suggested that spatial data types and functions—e.g., overlay, intersect, and touch—should be implemented in Ethereum. Goran et al. (2021) [11] then proposed a land information system (LIS) that integrates traditional LIS components—e.g., database server and web server—with the Ethereum public blockchain to allow geospatial functionalities.

3 System Design

3.1 Architecture

In our design, a blockchain client software would be installed in every government agency that wishes to participate in the system. Agencies will then connect their clients with each other to form a public permissioned blockchain network. When forming a network, one can choose a consensus algorithm that will secure the blockchain. Ethereum client software currently supports two algorithms, i.e., Ethash consensus, which is a proof-of-work algorithm, and Clique consensus, which is a proof-of-authority algorithm. In this case, we will use Clique consensus, because proof-of-authority is much less resource-intensive than proof-of-work. After the blockchain network is successfully established, a smart contract for managing the state land allocation will be deployed in the network. Each agency can interact with the smart contract by using the optional components, i.e., Listener, RDBMS, API Server, and Web Server.

Figure 1 (left) depicts the logical architecture of the system. Clients from each agency are connected to form a single logical entity that stores the blockchain and executes smart contracts. For Ethereum networks, such an entity is called Ethereum Virtual Machine (EVM). Figure 1 (right) depicts mandatory and optional components that each participant could choose to install. Components may be installed in a standalone computer or separated computers, depending on the need and limitation of the participant.

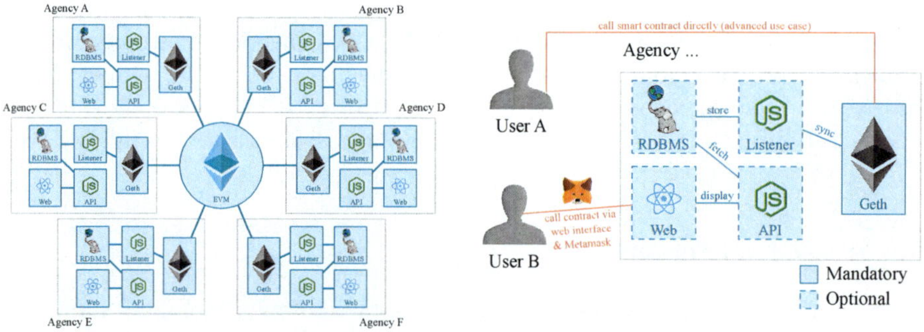

Fig. 1. Overall state land allocation system architecture (left). Components of the land allocation system (right).

- Blockchain client software is the main component that acts as a gateway for users in the agency to interact with the blockchain network. Any client software that conforms with the formal Ethereum specification can be used. In our case, we use Go Ethereum (Geth), an open-source client software implemented using the Go programming language.
- Listener is a component that extracts "events"—dispatched signals—emitted from the smart contract in the blockchain, then stores them in a relational database. When the listener starts, it will get and process every block since the last processed block. Afterward, the listener will always monitor the blockchain for new blocks to process. Listener can be written in any programming language that can use the Transmission Control Protocol and the Internet Protocol (TCP/IP) [12]. In our case, the language of choice is JavaScript. Since many Ethereum JavaScript Application Programming Interface (API) libraries, e.g., Web3.js and Ethers.js, already exist, the listener can be developed with ease.
- RDBMS is used to store data from the blockchain to improve data retrieval performance and perform sophisticated data queries, which could hardly be accomplished by using blockchains [13]. Due to data protection regulations that are effective in many countries, e.g., the General Data Protection Regulation (GDPR), data controllers and processors must prepare to delete personal data at some point [14]. Thus, sensitive data ought to be stored in the relational database rather than the blockchain, because data history in the blockchain can't be easily deleted. Another motive for using a relational database to store data is that some RDBMSs support geographic database types and functions. Since state land allocation data would typically be represented as polygons in Geographic Information Systems (GIS), its quality could be improved using various geographic tools provided by these RDBMSs. We use PostgreSQL as our RDBMS coupled with the PostGIS extension to add support for geographic functionality.
- API Server reformats and delivers data stored in the relational database to end users' clients—e.g., web, desktop, and mobile applications. Any state land location data that needs to be validated will also be sent to the API Server, which will then pass the data to the RDBMS to perform geospatial data validation. The API Server can be developed using any programming language. We use JavaScript to develop the API Server.
- Web Server is used to host static content. We mainly use it to host a web application that can greatly aid the users when interacting with the blockchain. Although users can technically use Blockchain client software to interact with the blockchain directly, it would be practically easier to use the web application.

3.2 Smart Contract

The state land allocation system relies on a single smart contract for user management, state land allocation registration, and state land activity registration. The smart contract's functionalities are described as follows.

- User management: The smart contract defines data structure for government agencies and their officers. An agency may have many officers. Officers are separated into two roles, i.e., administrators and operators. Any agency must have at least one administrator. Administrators will manage information about the agency—e.g., adding new users, managing state land allocation types that the agency oversees, and representing the agency when performing tasks that require voting among multiple agencies. Operators will use the smart contract to perform their assigned tasks.
- State land allocation registration: Each agency will have state land allocation types that it supervises. The smart contract will ensure that officers will only be able to manage state land allocation types that their agency supervises. Since Ethereum smart contracts don't support geographic data types, geographic data must be converted to a supported data type. We choose to store geographic data as Well-known Text (WKT) [15], a text representation for geometry data types originally defined by the Open Geospatial Consortium (OGC), which is supported by many software libraries.
- State land activity registration: Government agencies might launch campaigns or projects that target citizens who were allocated a state land—e.g., reforestation campaigns and agricultural subsidies. Storing data about these activities in the blockchain could prevent redundant campaigns in any state land and frauds, thus improving efficiency. Since such activities might target many types of state land, a state land activity registration record may refer to any state land in the blockchain.

3.3 Web Application

The web application will act as a client which allows end-users to interact with the API Server via their web browsers. We develop the web application as a single-page application (SPA) using React JavaScript library [16]. We also use Ethereum Provider JavaScript API [17] which allows end-users to interact with the blockchain via their web browsers by using MetaMask browser extension [18]. MetaMask is an extension that allows users to use their private key for transaction signing directly from their respective web browsers.

Since users use their private key to sign every transaction that would be sent to the blockchain and their authority is verified in the smart contract, additional authentication and authorization mechanisms aren't necessary. However, if some endpoints of the API Server are deemed to be classified, we could also implement a message-signing-based authentication mechanism in the API Server to allow users to authenticate by using their private key to sign a random string generated from the server, proving that they really own the private key [19]. This way, users can use their private key to authenticate both the blockchain and the API server without any need for additional credentials.

While most functionalities of the web application are developed based on the smart contract's functions, there are some functionalities that utilize the API Server, i.e., geospatial data validation and data listing.

- Geospatial data validation: If an officer submits a state land allocation to the system, the data would first be sent to the API Server in order to perform data validation. If there is no issue, a transaction confirmation window from MetaMask would then appear as depicted in Fig. 2. (left). If the officer confirms the transaction, his/her private key would then be used to sign the transaction before it is submitted to the blockchain. After the transaction reaches the blockchain, it will be validated by a specific group of nodes, i.e., "Signers" if the network operates in proof-of-authority mode or "Miners" if the network operates in proof-of-work mode.
- Data listing: When displaying a list of data, the web application will fetch data from the API Server for better performance. However, when displaying a single entity, the web application would directly fetch the data from the blockchain as depicted in Fig. 2 (right).

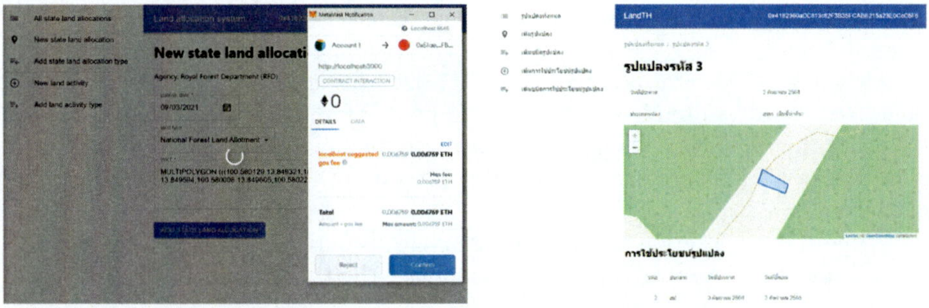

Fig. 2. Transaction confirmation window from MetaMask (left). Single entity data directly fetched from the blockchain (right).

4 Evaluation

The evaluation of our blockchain-based land allocation management system is performed in both performance aspect and feature aspect.

4.1 Performance Evaluation

We evaluated the performance of our system to ensure that it could handle transactions from multiple agencies. Data retrieval performance was not evaluated because each agency would retrieve data from their own blockchain node. Therefore, they can upgrade their nodes to improve data retrieval performance.

The prototype was installed on 4 virtual machines within a single physical machine as depicted in Fig. 3.

A simple script was written using JavaScript to continuously send transactions to one of the blockchain nodes. After every transaction was sent, we

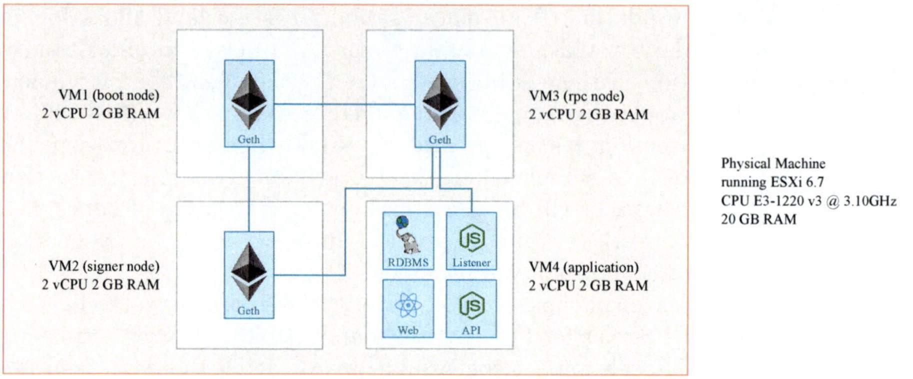

Fig. 3. System environment used in performance evaluation.

measured both the time used to send the transactions and the time used to create blocks.

Although the prototype can accept 417.62 transactions per second, transactions are actually stored in the transaction pool before being included in a block. By using the default settings, each block may contain 25 transactions and are created every 5 s. Therefore, the actual transaction throughput is 5 transactions per second.

While the transaction throughput is not considered high, it is sufficient for the use case. Moreover, it is possible to increase the throughput by changing the settings or optimizing the smart contract.

4.2 Feature Evaluation

We evaluate our prototype, representing a blockchain-based system, by comparing it with traditional information systems—systems that use centralized databases, in a scenario where each organization manages its own system. The comparison is presented in Table 1.

Since blockchains are shared ledgers that every participant can access, and all transactions are verified by multiple computer nodes before they are committed to the blockchain, data consistency can be guaranteed. Moreover, the blockchain data structure is designed to prevent malicious participants from tampering with it, resulting in an immutable data history that warrants data integrity.

Fundamentally, every participant that joins a blockchain network will have their own copy of the blockchain data. This is equivalent to having data backups at every participant's site, leading to high data resiliency. Although traditional information systems could also use disaster recovery sites to improve their data resiliency, the cost and operational complexity would usually be higher, and still couldn't reach the level of data resiliency that blockchains naturally provide.

Nevertheless, blockchain is still an emerging technology. Development tools for blockchain and developers who specialize in blockchain technology are still

Table 1. Comparison between blockchain-based systems and traditional information systems.

Characteristic	Blockchain-based	Traditional
Data consistency	Yes	No
Data quality	Yes	Yes
Data integrity	Yes	No
Data resiliency	High	Low
Development cost	High	Low
Deployment feasibility	Low	High
User interface (UI)	Yes	Yes
User experience (UX)	Yes	Yes

inadequate, resulting in higher development cost. Furthermore, deploying a blockchain network involves multiple government agencies. Thereby, mechanisms for formalizing partnerships among agencies, e.g., memoranda of understanding (MOUs), are also required.

5 Deployment Considerations

Compared with traditional information systems, state land allocation system—which is blockchain-based—have some additional factors to be considered before deployment as follows.

User Accounts Generation. Alike traditional information systems, blockchain-based systems also have mechanisms to authenticate users. However, user accounts in blockchain are basically key pairs generated using public key algorithms. Users can generate their own key pairs using any software that follows the specification of the blockchain. One of the key pairs, namely the private key, will be used to digitally sign transactions in the blockchain, therefore it must be kept secret. For Ethereum blockchains, since all transactions in the network are subject to fees, some cryptocurrencies must also be distributed to users. In addition, every user who wishes to interact with the system via web browsers will have to install MetaMask browser extension.

Genesis Block Creation. Every blockchain starts with the genesis block. By default, Ethereum clients will use the Mainnet's genesis block. But in our case, we will create an Ethereum private network to avoid transaction fees, so a genesis block is necessary. Some settings, e.g., consensus algorithm to use and initial allocation of cryptocurrency—which is required for transactions in the blockchain, can only be configured in the genesis block.

Smart Contract Approval. Since smart contracts can't be modified after they were deployed, agencies must carefully review the smart contract—to ensure that it would perform as expected—prior to usage. Some rules—e.g., the number of votes required before a new agency can participate in the network—can be enforced automatically by the smart contract.

6 Future Works

Though the state land allocation system could successfully act as a single source of truth for multiple government agencies. There are still many interesting problems for future works.

6.1 Blockchain-Level Geospatial Data Validation

Due to technological limitations, the current system can only perform simple data validation—such as enforcing required field and numbers comparison—at the blockchain level. Complicated data validation, in this case, geospatial data validation, can only be done at the application level—before data enter the blockchain. This may cause conflicts in some cases—e.g., multiple agencies submit overlapping land allocations at the same time—because of the delay before the application caches the most recent data from the blockchain. If geospatial data validation could be performed at the blockchain level, overall data consistency would be further improved.

6.2 Smart Contracts Ecosystem

For simplicity, the current smart contract performs multiple tasks, e.g., user management, state land allocation registration, and state land activity registration. Since Ethereum smart contracts can call functions of other smart contracts, these tasks could be separated into multiple smart contracts. This would make each smart contract simpler and more modular. In addition, government agencies could incrementally add new functions by creating new smart contracts which might interact with old smart contracts.

7 Conclusion

This paper proposed a prototype of a blockchain-based system that was designed to facilitate collaboration in state land allocation data management among multiple agencies. It also discussed the architecture and essential components of the system, evaluation of the system when compared with traditional information systems, as well as some factors that need to be considered when deploying a blockchain-based system. Ethereum client software was used to create a proof-of-authority blockchain network. The system also used a relational database, PostgreSQL with PostGIS extension, to cache blockchain data, store sensitive data and perform geospatial data validation.

References

1. Calvery, J.S.: Statement of Jennifer Shasky Calvery, Director Financial Crimes Enforcement Network United States Department of the Treasury. Financial Crimes Enforcement Network, Vienna, Virginia, United States (2013)
2. Pahl, C., El Ioini, N., Helmer, S.: A decision framework for blockchain platforms for IoT and edge computing (2018). https://doi.org/10.5220/0006688601050113
3. Androulaki, E., et al.: Hyperledger fabric: a distributed operating system for permissioned blockchains. In: Proceedings of the Thirteenth EuroSys conference, pp. 1-15 (2018)
4. Brown, R.G., Carlyle, J., Grigg, I., Hearn, M.: Corda: An Introduction. R3 CEV, August 1, 15 (2016)
5. Wood, G.: Ethereum: a secure decentralised generalised transaction ledger (2014). https://gavwood.com/paper.pdf. Accessed 06 Feb 2022
6. Chang, K.T.: Geographic Information System, pp. 1-10. American Cancer So- ciety (2019). https://doi.org/10.1002/9781118786352.wbieg0152.pub2. Accessed 06 Feb 2022
7. Postgis Developers: Postgis - spatial and geographic objects for postgresql. https://postgis.net. Accessed 06 Feb 2022
8. PostgreSQL Global Development Group: Postgresql. https://www.postgresql.org. Accessed 06 Feb 2022
9. Mendi, A.F., Çabuk, A.: Blockchain applications in geographical information systems. Photogram. Eng. Remote Sens. J. **86**(1), 5–10 (2020)
10. Farnaghi, M., Mansourian, A.: Blockchain, an enabling technology for transparent and accountable decentralized public participatory GIS. Cities **105**, 102,850 (2020)
11. Sladi'c, G., Milosavljevi'c, B., Nikoli'c, S., Sladi'c, D., Radulovi'c, A.: A blockchain solution for securing real property transactions: a case study for Serbia. ISPRS Int. Jo. Geo-Inf. **10**(1), 35(2021). https://doi.org/10.3390/ijgi10010035
12. Ethereum community: Programming languages. https://ethereum.org/en/developers/docs/programming-languages. Accessed 06 Feb 2022
13. Chitti, P., Murkin, J., Chitchyan, R.: Data management: relational vs blockchain databases. In: Proper, H., Stirna, J. (eds) Advanced Information Systems Engineering Workshops. CAiSE 2019. Lecture Notes in Business Information Processing, vol. 349. Springer, Cham (2019). https://doi.org/10.1007/978-3-030-20948-3_17
14. Tikkinen-Piri, C., Rohunen, A., Markkula, J.: Eu general data protection regulation: changes and implications for personal data collecting companies. Comput. Law Secur. Rep. J. **34**(1), 134–153 (2018)
15. Open Geospatial Consortium: Simple feature access - part 1: Common architecture. https://www.ogc.org/standards/sfa. Accessed 06 Feb 2022
16. React. https://reactjs.org. Accessed 06 Feb 2022
17. Vogelsteller, F., Ghods, R., Maia, V., Garreau, M., Marks, E.: Eip-1193: Ethereum provider javascript api (2018). https://eips.ethereum.org/EIPS/eip-1193. Accessed 06 Feb 2022
18. Metamask - a crypto wallet & gateway to blockchain apps. https://metamask.io. Accessed 06 Feb 2022
19. Martiny, A.: Metamask tutorial: One-click login with blockchain made easy (2018). https://www.toptal.com/ethereum/one-click-login-flows-a-metamask-tutorial. Accessed 06 Feb 2022

Modified Ant System with Threshold for the Vehicle Routing Problem

Petcharat Rungwachira and Arit Thammano(✉)

Computational Intelligence Laboratory, Faculty of Information Technology, King Mongkut's Institute of Technology Ladkrabang, Bangkok 10520, Thailand
59606004@kmitl.ac.th, arit@it.kmitl.ac.th

Abstract. Vehicle routing problem is a combinatorial optimization problem. Ant system, one of the most popular algorithms for solving the combinatorial optimization problem, is inspired by the ant foraging behavior. However, Ant system has the problem of premature convergence and easily getting trapped into local optimum. This paper proposes the modified ant system with threshold. In the beginning, the proposed algorithm randomly generates the initial population. To obtain a more diverse population, the transition probability and the threshold are used to determine which city should be next in the path. Moreover, the improved pheromone updating process is introduced to help reducing the rate of convergence. Three local searches, swap, insert, and reverse, are used in the proposed algorithm to prevent getting trapped in a local optimum. The datasets used in this research were taken from TSPLIB, the standard benchmark datasets for TSP. The performance of the algorithm was compared with that of the other 3 algorithms: GA-PSO-ACO, FOGS-ACO, and Hybrid VNS. The results on 20 datasets showed that the proposed method outperformed or at least at par with the others on 15 datasets. On the other hand, there was one dataset, rd100, on which the proposed algorithm was better than the best-known solution. According to the result, our modified ant system with threshold was very effective for the small- and medium-sized vehicle routing datasets.

Keywords: Vehicle routing problem · Combinatorial optimization · Ant system

1 Introduction

At present, the purchase of products has been changed from shopping at stores to be an online purchase. Especially, the Covid-19 outbreak causes the online purchase to become more popular. Therefore, an efficient transportation routing is needed to reduce costs and increase customer satisfaction. Vehicle Routing Problem (VRP) is a combinatorial optimization problem with the objective of finding the shortest route to traverse all the cities without repeating any city. The shortest route does not only reduce the use of oil, which is a limited resource, but also helps to reduce air pollution as well.

Although many researches have been studied about the problem of VRP, there is still a lot of room to improve. Therefore, many interested algorithms have been introduced.

P. Meesad et al. (Eds.): IC2IT 2022, LNNS 453, pp. 22–31, 2022.
https://doi.org/10.1007/978-3-030-99948-3_3

The most common type of algorithms for solving this problem is metaheuristic. Although the result of the metaheuristic may not be the optimal solution, it is the best solution within a reasonable time. The popular algorithms for solving the problem are Ant Colony Optimization (ACO), Genetic algorithm (GA), and Particle Swarm Optimization (PSO).

A hybrid approach for solving Traveling Salesman Problem (TSP) was proposed by Mahi et al. [1]. The approach consisted of Particle Swarm Optimization (PSO), Ant Colony Optimization (ACO), and 3-Opt. ACO was used to generate the route. PSO was used to determine the values of α and β, which were the weight parameters in the ACO. 3-Opt was used as a local search. The experiments were performed on different number of ants. The result showed that the number of ants significantly affected the performance of the hybrid.

A Discrete Symbiotic Organism Search (DSOS) was presented by Ezugwu and Adewumi [2]. It was a discrete version of SOS for solving TSP. Three mutation operators, swap, insertion, and inversion, were added to the proposed algorithm to increase the diversity of a new population. The datasets used for evaluating the algorithm were from TSPLIB. The experimental results showed that their proposed algorithm performed better than the other algorithms.

A hybrid algorithm which combined PSO and Artificial Bee Colony (ABC) algorithms was proposed by Sedighizadeh and Mazaheripour [3]. It emphasized on precedence constraints for optimizing multi-objective VRP. PSO and ABC were used to construct the good routes in parallel. The result showed that the proposed algorithm was able to locate the route more quickly than the other algorithms.

A discrete firefly algorithm was proposed by Osaba et al. [4] for solving Rich Vehicle Routing Problem (RVRP). Discrete firefly algorithm was used in this algorithm to manage both pickup and delivery of newspapers. The experimental results on 15 datasets of real-world locations showed that the discrete firefly algorithm outperformed the others two algorithms.

HAFA, a hybrid of ACO and FA algorithms, was presented by Goel and Maini [5] for solving VRP. ACO was used as the basis of the algorithm while FA with a new distance measure technique was used to explore new solutions in a search space. Pheromone shaking was used in this hybrid algorithm for adjusting pheromone levels to prevent getting trapped in the local optimum. Experimental results of this algorithm showed that it performed well on small- and medium-sized datasets with less than 200 customers.

CluVRP, a novel two-level optimization approach for solving clustered VRP, was presented by Pop et al. [6]. CluVRP divided the problem into upper and lower sub-problems. GA was used to solve the upper sub-problem for the global route to visit the cluster of customers, while the lower sub-problem was used to determine how to visit all customer in the cluster by using a Concorde TSP solver.

A hybrid algorithm that decomposed a large-scale TSP into small-scale one was presented by Liao and Liu [7]. The customers were clustered by Density Peak Clustering. The subroute that connected all customer in each cluster was created by ACO. Then the global route was constructed by iteratively connecting two nearest subroute together until there was only one route remaining. Finally, the global route was optimized by k-opt algorithm.

A hybrid two stage Sweep Algorithm (SA) with greedy search for solving CVRP was presented by Chen et al. [8]. SA was used in the algorithm to cluster all customers into groups. Then the re-cluster was done by adding customers from the adjacent clusters, depending on the capacity of the vehicle, in order to avoid the solution getting trapped in a local minimum. Finally, a greedy search was used to find the shortest route for each vehicle.

This paper is organized in 5 sections as follows: Sect. 1 is the introduction; In Sect. 2, some definitions and background concepts are described; In Sect. 3, the proposed method is discussed; The experimental results are shown and discussed in Sect. 4; Finally, Sect. 5 is the conclusion.

2 Background

This section briefly discusses about the objective function of VRP and the original Ant System.

2.1 Objective Function of VRP

The objective function of VRP is the total distance traversed by the vehicle, Z, as shown in (1).

$$Z = \sum_{i=1}^{n} \sum_{j=1, j \neq i}^{n} \eta_{ij} \tag{1}$$

where n is the number of cities or customers, and η_{ij} is the distance between city i and city j.

2.2 Ant System (AS)

AS is inspired by the foraging behavior of ants which can find the shortest path from the food source to their colony. Ants use their pheromone to inform about their journey to other ants. Pheromone level on the edge increases with the number of ants passing through that edge. On the other hand, if no ant passes through the edge, the pheromone level on that edge will be decreased over time. When using AS to solve VRP, each ant represents a candidate solution to a problem. In general, the number of ants is assigned equal to the number of cities.

At the beginning of the algorithm, each ant is assigned a different starting city to increase path diversity. Then the city J with the highest transition probability from city i is selected. The transition probability P_{ij}^{k} depends on the distance and the pheromone on the edge ij as shown in (2).

$$P_{ij}^{k} = \frac{(\tau_{ij})^{\alpha}(\eta_{ij})^{\beta}}{\sum_{l \in p}(\tau_{il})^{\alpha}(\eta_{il})^{\beta}} \tag{2}$$

$$J = \arg\max_{j}(P_{ij}^{k}) \tag{3}$$

where P_{ij}^{k} represents the transition probability from city i to city j of the ant k; τ_{ij} is the pheromone level on the edge between city i and city j; η_{ij} is the distance between city i and city j; α and β are the weight of pheromone and distance, respectively. The process of selecting the next city continues until the path is completed. The total distance of the path is then calculated. After all paths are completed, these paths are compared to find the shortest one.

Finally, the pheromone level on each edge is updated with the information from all paths. The equation for updating pheromone is shown below.

$$\tau_{ij}^{'} = (1-\rho)\tau_{ij} + \sum_{k=1}^{K} \Delta\tau_{ij}^{k} \tag{4}$$

$$\Delta\tau_{ij}^{k} = \frac{1}{Z_k} \tag{5}$$

where ρ represents the evaporation rate; Z_k is the total distance of the path traversed by the ant k.

3 Our Proposed Method

Modified Ant System with Threshold (M-Ant-T) and the local search operators for solving VRP are described in this section. The flowchart of our M-Ant-T is shown in Fig. 1 and the detail of the algorithm is described as follows.

3.1 Modified Ant System with Threshold

MMAS [9], which is a modified version of AS, is used as the core of this proposed algorithm.

Step 1 Initialize parameters
In this first step, the initial values of parameters are assigned: the population size (K), the number of cities (n), the maximum number of ants in *G-Set*, the evaporation rate (ρ), the initial pheromone ($\tau(0)$), the maximum number of iterations, and α and β.

Step 2 Generate the initial population
The initial population is randomly generated to increase the diversity of the population. The number of ants in the population should not be less than the number of cities.

Then the fitness value of each ant is calculated. Note that the fittest ant is the one with the lowest distance. Next, the best n ants are copied into the *G-Set*.

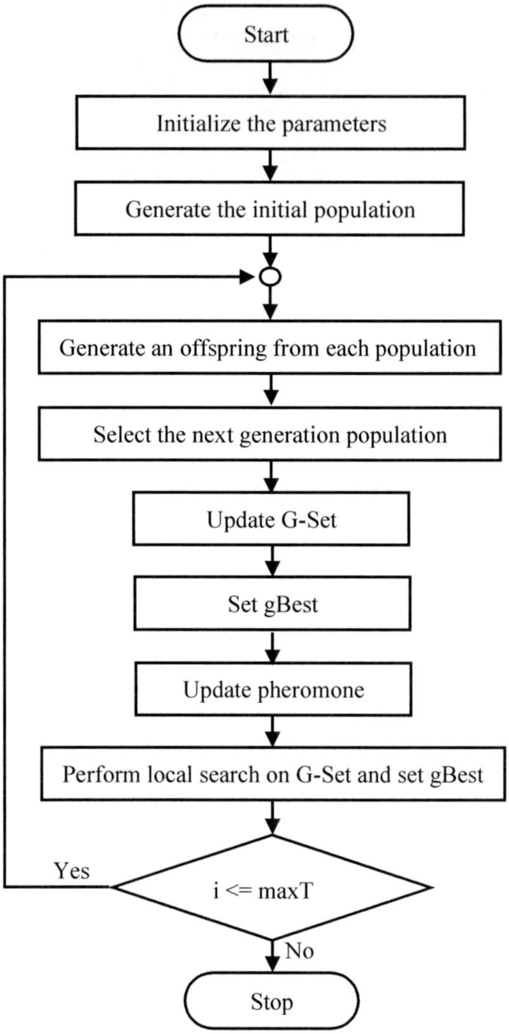

Fig. 1. Flowchart of the M-Ant-T.

Step 3 Loop until the maximum number of iterations is reached

- Generate the offspring.
  - A position in the parent solution is randomly selected. Then the subpath starting from the selected position to the end of the solution is used as the starting subpath of the offspring.
  - The next city in the offspring is selected one by one until the offspring is completed. To select the next city, the transition probabilities from the last city in the subpath to all unassigned cities are calculated. Then the city J with the highest transition

probability is determined. If $P_{iJ}^{k} \leq Threshold$, the city J is selected to be the next city. Otherwise, the city with the second-highest transition probability is selected to be the next city. Threshold is a random number; it is introduced here to produce more diverse offspring.

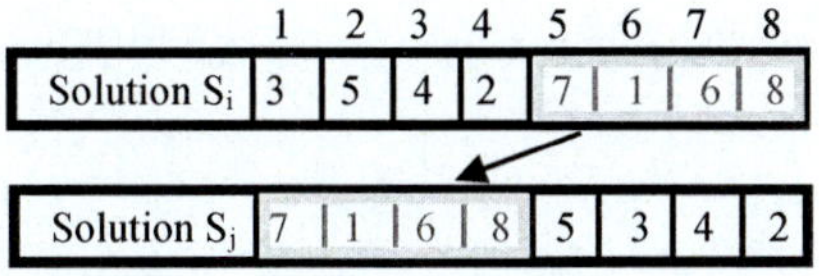

Fig. 2. Generate an offspring.

For example, as illustrated in Fig. 2, if S_i is a parent solution and 5 is a randomly selected position, the subpath will be 7-1-6-8. Therfore, the subpath 7-1-6-8 is used as the starting subpath of S_j. The remaining cities, 5, 3, 4 and 2, are selected one by one according to the above described process.

- The total distance of each offspring is calculated according to (1).
- The n best solutions from the combined pool of parents and offspring are selected to be the next generation population.
  - The selection of the next generation population is divided into two parts: 70% are selected in order of their fitness values, and 30% are randomly selected from the rest.
  - If any solution in the next generation population is better than the worst solution in *G-Set*, *G-Set* will be updated.
  - The solution with the least total distance is set to be *gBest*, global best.
- The pheromone on each edge is updated based on the s^{best} solutions by using (6) and (7).

$$\tau_{ij}^{'} = (1 - \rho)\tau_{ij} + \Delta\tau_{ij}^{best} \tag{6}$$

$$\Delta\tau_{ij}^{best} = \frac{1}{f\left(s^{best}\right)} \tag{7}$$

  where s^{best} may be either *gBest*, the best or the second-best solutions in each iteration.
- Local search is performed on the solutions in G-Set.

Step 4 Return ***gBest.***

3.2 Local Search

The local search used in our proposed algorithm, M-Ant-T, are Swap, Insert, and Reverse operators.

1. Swap operator
   Swap operator randomly selects two positions, *i* and *j*. Then the two selected cities are swapped (Fig. 3).

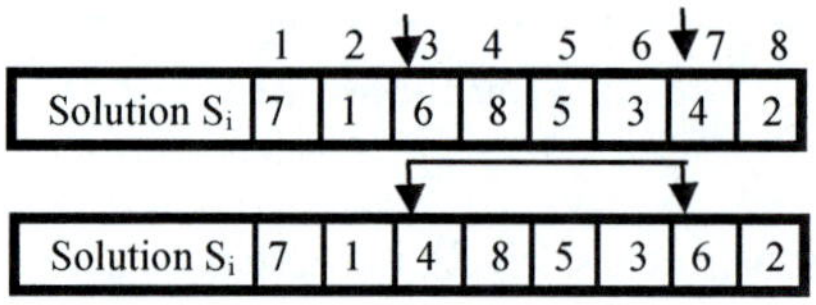

Fig. 3. Swap operator.

2. Insert operator
   Insert operator randomly selects two positions, *i* and *j*. Then the city in the position *j* is inserted in front of the position *i* (Fig. 4).

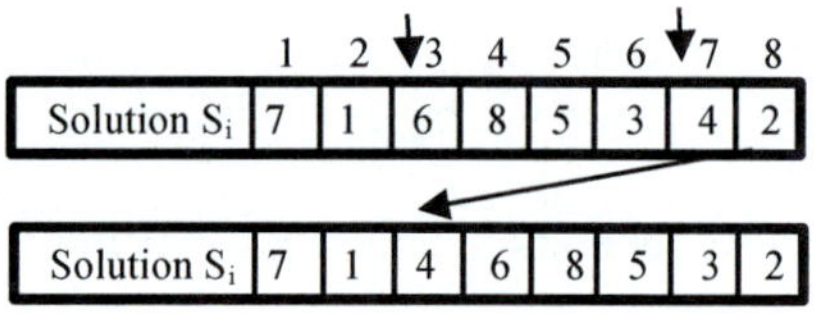

Fig. 4. Insert operator.

3. Reverse operator
   Reverse operator randomly selects two positions, *i* and *j*. Then the sequence of the cities between these two positions are flipped (Fig. 5).

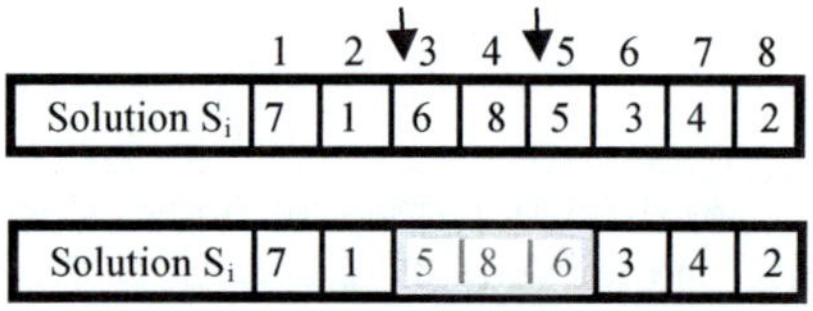

Fig. 5. Reverse operator.

4 Experimental Results

The standard benchmark datasets used for evaluating the performance of our algorithm were taken from TSPLIB [10]. They are available at http://comopt.ifi.uni-heidelberg.de/software/TSPLIB95/tsp/. Twenty small- and medium-sized datasets were used in the experiment, namely att48, eil51, berlin52, st70, pr76, eil76, rat99, kroA100, kroB100, kroC100, kroD100, kroE100, rd100, eil101, lin105, pr124, ch130, ch150, kroA150, and kroA200.

The initial values of the parameters used in this experiment are shown in Table 1. The evaporation rate, ρ, was set to between 0.05–0.5. The initial pheromone, $\tau(0)$, was set to be 3. The weight parameters for pheromone level and distance, α and β, were set to be 1 and 2, respectively. The number of iterations was set to be 5,000 for small-sized datasets, and 10,000 for medium-sized datasets.

Table 1. The initial parameter values.

Parameters	Values
ρ	0.05–0.5
$\tau(0)$	3
α	1
β	2

The comparison between M-Ant-T and three state-of-the-art algorithms were shown in Table 2. GA-PSO-ACO was proposed by Deng et al. [11], FOGS-ACO was proposed by Saenphon et al. [12], and Hybrid VNS was proposed by Hore et al. [13].

When comparing between M-Ant-T and GA-PSO-ACO on 15 datasets, M-Ant-T was better than or equal to the best of GA-PSO-ACO on 11 datasets. When comparing between M-Ant-T and FOGS-ACO on 10 datasets, M-Ant-T was better than or equal to the best of FOGS-ACO on 9 datasets. In addition, when comparing between M-Ant-T and Hybrid VNS on 19 datasets, M-Ant-T was better than or equal to the best of Hybrid VNS on 16 datasets. However, on rd100 dataset, M-Ant-T was better than the best-known solution.

Table 2. The comparison between GA-PSO-ACO, FOGS-ACO, Hybrid VNS, and M-Ant-T.

	Instance	Best-Known Solution	GA-PSO-ACO	FOGS-ACO	Hybrid VNS	M-Ant-T
1	att48	33522	33524	33561	-	**33522**
2	eil51	426	**426**	**426**	428.98	**426**
3	berlin52	7542	7544.37	7546.62	7544.36	**7542**
4	st70	675	679.6	678.93	677.11	**675**

(continued)

Table 2. (*continued*)

	Instance	Best-Known Solution	GA-PSO-ACO	FOGS-ACO	Hybrid VNS	M-Ant-T
5	pr76	108159	109206	108864	**108159**	**108159**
6	eil76	538	545.39	546.83	545.39	**541**
7	rat99	1211	1218	-	1240.38	**1212**
8	kroA100	21282	-	21414	21618.2	**21282**
9	kroB100	22141	-	-	**22139.07**	22179
10	kroC100	20749	-	-	20750.76	**20749**
11	kroD100	21294	21394	-	21294.29	**21294**
12	kroE100	22068	-	-	22174.6	**22073**
13	rd100	7910	7936	7919	7910.4	**7906**
14	eil101	629	**633.07**	633.4	642.31	634
15	lin105	14379	**14397**	-	14383	**14379**
16	pr124	59030	59051	-	**59030.74**	59076
17	ch130	6110	**6121.15**	-	6140.66	6174
18	ch150	6528	-	-	6639.52	**6528**
19	kroA150	26524	**26676**	-	26943.31	26686
20	kroA200	29368	29731	29717	30300.56	**29617**

5 Conclusion

In this paper, M-Ant-T is presented. It was used to modify the updated pheromone of ant system and then the threshold was used to create different paths. The evaluation was performed on 20 small- and medium-sized datasets of VRP. When comparing M-Ant-T with the other 3 state-of-the-art algorithms, GA-PSO-ACO, FOGS-ACO and Hybrid VNS, the results showed that the proposed algorithm was better than or equal to those algorithms on 15 datasets. In addition, on one of these datasets, rd100, M-Ant-T performed even better than the best-known solution. However, this paper emphasized only on small- and medium-sized datasets. In our future work, we plan to improve the algorithm to be more efficient for medium- and large-sized datasets.

References

1. Mahi, M., et al.: A new hybrid method based on particle swarm optimization, ant colony optimization and 3-opt algorithms for traveling salesman problem. Appl. Soft Comput. **30**, 484–490 (2015)
2. Ezugwu, A.E.-S., Adewumi, A.O.: Discrete symbiotic organisms search algorithm for travelling salesman problem. Expert Syst. App. **87**, 70–78 (2017)

3. Sedighizadeh, D., Mazaheripour, H.: Optimization of multi objective vehicle routing problem using a new hybrid algorithm base on particle swarm optimization and artificial bee colony algorithm considering precedence constraints. Alexamdria Eng. J. **57**(4), 2225–2239 (2017)
4. Osaba, E., et al.: A discrete firefly algorithm to solve a rich vehicle routing problem modeling a newspaper distribution system with recycling policy. Soft. Comput. **21**, 5295–5308 (2017)
5. Goel, R., Maini, R.: A hybrid of ant colony and firefly algorithm (HAFA) for solving vehicle routing problems. J. Comput. Sci. **25**, 28–37 (2018)
6. Pop, P.C., et al.: A novel two-level optimization approach for clustered vehicle routing problem. Comput. Ind. Eng. **115**, 304–318 (2018)
7. Liao, E., Liu, C.: A hierarchical algorithm based on density peaks clustering and ant colony optimization for traveling salesman problem. IEEE Access **6**, 38921–38933 (2018)
8. Chen, M.H., et al.: A hybrid two-stage sweep algorithm for capacitated vehicle routing problem. In: International Conference on Control, Automation and Robotics 2015, Automation and Robotics (ICCAR), IEEE (2015)
9. Zao, G., et al.: A modified max-min ant system for vehicle routing problems. In: Proceedings of the 4th International Conference on Wireless Communications, Networking and Mobile Computing (2008)
10. TSPLIB. http://comopt.ifi.uni-heidelberg.de/software/TSPLIB95/tsp/. Accessed 30 Dec 2021
11. Deng, W., et al.: A novel two-stage hybrid swarm intelligence optimization. Soft. Comput. **16**, 1707–1722 (2012)
12. Saenphon, T., et al.: Combining new fast opposite gradient search with ant colony optimization for solving travelling salesman problem. Eng. Appl. Artif. Intell. **35**, 324–334 (2014)
13. Hore, S., et al.: Improving variable neighborhood search to solve the travelling salesman problem. Appl. Soft Comput. **68**, 83–91 (2018)

Discovery of Business Model Transform in Commerce Using LSA Approach with Firm Information Disclosure

Sompong Promsa-ad and Nichnan Kittiphattanabawon(✉)

School of Informatics, Walailak University, 222 Thasala District, 80161 Nakhonsithammarat, Thailand
knichcha@wu.ac.th

Abstract. In order to survive and grow in rapidly changing settings, incumbents and established commerce enterprises must adapt their business models. However, changing a business model is a difficult task because there is no best practice provided. In our study, we wanted to learn from the best-performing firms in order to find such an effective solution. We used data from annual reports of enterprises in the Stock Exchange of Thailand's commerce business sector to perform Latent Semantic Analysis (LSA). Tobin's Q ratio and CAGR were used to select successful and fast-growing businesses, resulting in a final sample of 14 prominent businesses. Our TF-IDF calculations yielded 5,596 words after the text preparation step. Our analyses indicated that our corpus contained three latent topics. We may conclude, based on the key terms for each topic, that developing online sales channels alongside traditional business forms is an essential business model transformation approach for companies looking to survive and grow in the digital age.

Keywords: Business model transform · Latent semantic analysis · Information disclosure

1 Introduction

Retail and wholesale enterprises that provide finished goods to consumers are known as commerce businesses. They might run their business from a physical location or over the internet. The retail environment is more competitive than it has ever been. Most incumbent retailers are encountering new sorts of competitive forces as digital-related technologies advance. For example, on-demand fulfillment is changing how and where businesses handle goods [1]. In order to survive and grow in such a quickly changing environment, incumbents and established retailers must adapt their business models.

The emergence of new players, particularly online marketplaces, is driving the need to adapt into new retail formats. These marketplaces excel at growing market demand for highly specialized products. Most digital retail platforms earn from each transaction by exploiting market inefficiencies without carrying any inventory [2]. The effect of an expansion of the online selling channel is considered a phenomenal. Global e-commerce sales were 2.3 trillion dollars in 2017 and an e-retail revenues are forecasted to increase

P. Meesad et al. (Eds.): IC2IT 2022, LNNS 453, pp. 32–42, 2022.
https://doi.org/10.1007/978-3-030-99948-3_4

to 4.88 trillion dollars in 2021 [3]. Despite its expanding importance, changing business models is a difficult task. Because every corporation will keep it a well-guarded secret to protect their competitive advantages, the most effective solution to the problem is also inaccessible. However, the best practice of the issue may reside in firm information disclosure since the signaling theory suggest that management of best-performing companies tend to disclose more information to the investors directly or indirectly intending to signal their prosperity [4].

Extracting the guideline for business model transform using text from a firm's disclosure or annual report is a difficult task, as one will be overwhelmed by a big corpus of textual data. However, the stated aim is now attainable because to the existence of learning algorithms that can cope with high-dimensional inputs such as nonconventional unstructured information such as textual data. Machine learning techniques are increasingly being used to research finance and corporate dynamics. For example, in [5], the authors utilize text and data mining to study the association between annual report text patterns and company sales performance. Text analytics can also be used to solve problems like financial statement fraud detection, as proposed by [6], and topic extraction for banks, as proposed by [7].

In this study, our main objective is to develop a solution or best practice for business model transformation by learning from successful or fast-growing companies. In order to find how those top-performers adapt their business models to the changing business environment and competitive landscape, we use Latent Semantic Analysis with information obtained from annual reports of companies belonging to commerce business sector of the Stock Exchange of Thailand. The sector is regarded as one of the most disrupted areas in the digital era. To our knowledge, no other study has taken the same approach to the problem. Furthermore, no research in these areas has been undertaken utilizing the Thai language version of the firm's annual report.

The remainder of the article is organized as follows. Section 2 summarizes the literature on extracting hidden issues from firm information disclosure. Section 3 explains the dataset and data preprocessing procedure. The experiments and key findings are summarized in Sect. 4. The findings are discussed in Sect. 5, and the conclusion is shown in the final section.

2 Extracting Hidden Topics from Information Disclosure

2.1 Information Disclosure

Investors are believed to make decisions based on information voluntarily disclosed by corporations. This data is deemed useful because signaling theory predicts that top-performing companies' managements are more likely to divulge more information to investors, either directly or indirectly, in order to signal their superiority and, as a result, attract capital [4]. The annual report is usually one of the most important sources of information disclosure for publicly traded companies. The report is a public document that depicts a company's financial situation and operational performance during a fiscal year. In addition, the report usually contains the company's fundamental business content, as well as its competitiveness, risk factors, and strategic options, among other things. It is expected that when a corporation makes significant strategic changes, the changes will be reported in the annual report [5, 8].

Previous research on the influence of information disclosure on corporate value in the financial and accounting literature largely examined voluntary, environmental, sustainability, intangibles, and other sorts of data [4]. As digital transformation becomes an inescapable option, recent research has examined how digital transformation disclosure influences company performance. In most studies, the frequency of relevant concepts is used as a proxy for the level of information disclosure or digital engagement. The findings were inconclusive. While [4, 8, 9] discovered a link between the extent of digital transformation information disclosure and firm performance, [10] discovered that, while there is a link between digital practices and firm performance, it is rather weak.

Despite the fact that the majority of information disclosure research focuses on the influence of disclosure level on firm performance, researchers and practitioners are becoming increasingly interested in a subset of literature that seeks to extract hidden useful messages from disclosure. However, deducing latent meaning from text is a truly challenging endeavor because one is confronted with a large corpus of textual data. As a result, researchers are moving to more automated methodologies based on machine learning techniques, such as Latent Semantic Analysis (LSA) and Latent Dirichlet Allocation (LDA). Corporate social responsibility (CSR) and business strategy studies have both utilized these methodologies [11, 12].

2.2 Latent Semantic Analysis

Latent Semantic Analysis (LSA) was first introduced as Latent Semantic Indexing (LSI) with the purpose of indexing text. LSA extends the notion beyond indexing to all analytic applications [13]. It is one of the most often used tools for document summarizing. The core premise of LSA is that there is a latent structure among terms in every document that are contextually connected and should thus be correlated in the same singular space [14]. As a result, it is to discover the hidden topics represented by the document or text. The discovered hidden themes are then used to group documents that are similar.

In LSA, the arithmetic utilized to discover the meaning of words is known as singular value decomposition (SVD). Some text preprocessing processes, such as word tokenization, stop word removal, and stemming, are required before executing LSA. The expected outcome of this procedure is a document-term representation, which will be used as an analysis input. In this regard, the most commonly used weighing method is term frequency-inverse document frequency (TF-IDF). An exceedingly sparse matrix will be converted into a small number of latent dimensions after SVD calculation. The model also determines the topic to which each document belongs, as well as the relevant key-words connected with each topic [13].

3 Dataset and Data Preparation

3.1 Dataset

The goal of this study is to find the best practice of business model transformation by learning from successful or fast-growing firms. These top-performing organizations' solutions can help enterprises seeking for a blueprint for business model adaption. Therefore, we select samples based on their growth prospects. We use firms in commerce business sector from the Stock Exchange of Thailand (SET) as samples because the sector is considered as one of the most impacted industries by digital disruption. Commerce firms, according to SET, are those that provide finished goods to consumers in the form of retail and wholesale. They can operate in a physical store, such as shops, department stores, convenience stores, discount stores, and superstores, or online. There are 30 businesses in this industry listed in SET as of December 2021.

In our selection procedure, we consider two aspects. Financial data provided from SETSMART is used to evaluate these factors. To begin, we use Tobin's Q to assess firms' growth prospects, using the formula as employed in [15]. Tobin's Q is a widely used indicator of a company's intangible value, which represents its growth prospects. Tobin's Q ratio is calculated as the market value of equity plus total assets minus the book value of equity, divided by book value of total assets [16]. Companies must have a Tobin's Q greater than 1 for five consecutive years between 2016 and 2020 to be selected for the sample.

The second criterion is the performance of firms in terms of sales. To be considered for the sample, firms must have a cumulative annual growth rate (CAGR) of sales of at least 20% during the last five years ending in 2020. With that kind of growth, they will be classified as fast-growing companies [17]. To be considered for the sample, the company must meet at least one of the requirements listed. After the screening process, the final sample consisted of 14 firms.

The primary source of data for this study is the annual reports for the year 2020 of selected companies, which may be obtained on their websites. For the analysis, we use the whole text of the firms' business descriptions, as well as the market and competition sections of the reports. The present operational model of the company is discussed in these sections. It also illustrates how the business responds to changing business conditions, competitive landscapes, and industry trends. As a result, we anticipate that the sections will be able to provide some insight into how a company's business model evolves.

3.2 Data Preprocessing

Natural language processing techniques are used to refine the text prior to LSA analysis. We use PyThaiNLP, a Python package for text processing and linguistic analysis, because our data contains Thai text. This is a text preprocessing tool comparable to NLTK, but with a concentration on Thai. Thai words, English words, figures, special characters, and other items can be found in the annual reports. Therefore, we do data preprocessing activities such as word segmentation, punctuation removal, stop-word removal, and whitespace removal. The process includes changing English uppercase letters to lowercase. Then we use the term frequency-inverse document frequency (TF-IDF) approach to extract terms. The TF-IDF is thought to be a good method to LSA [13]. It enables us to select representative terms that reflect the content's heart. When a term appears frequently in a few documents, the TF-IDF has a high value, and when it appears in the majority of documents, the TF-IDF has a low value. We acquired 5,596 words as a result of our TF-IDF computations based on 14 documents in our corpus.

4 Experiments

4.1 Determining Optimal Number of Topics

To identify the appropriate number of topics to be included in our LSA study, we use the approach described by [13]. To begin, we perform TruncatedSVD and plot the outcome as a scree plot. Then we search for an elbow point to determine the number of factors. As depicted in Fig. 1, the horizontal axis represents the number of factors and the vertical axis represents the percentage of variance explained by each of the selected components. Figure 1 shows that an elbow point with a k = 2 could exist. This means that little or no further information is given beyond the chosen k. The number of concepts utilized to project the data to the reduced k-dimensional concept in this study is 3 because the index starts at 0.

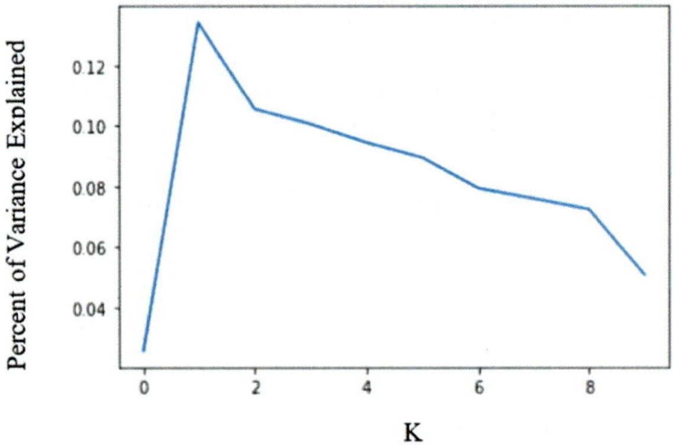

Fig. 1. Percent of variance explained with increasing *k* value.

4.2 Latent Topics and Relevant Key Terms

To obtain the latent concepts and relevant key terms, we employ TruncatedSVD with three components. Figure 2 depicts the first latent topic, which is made up of meaningful words. These are the terms to which the model gave the highest weight. As observed, words relating to the fundamental operations of most retail businesses, such as product (สินค้า), client (ลูกค้า), branch (สาขา), sell (จำหน่าย), and so on, dominate the top 20 phrases. It is worth noting that this dimension includes the key-word online (ออนไลน์), which is a term that relates to a digital transformation.

The top 20 keywords in the second and third dimensions demonstrate the least amount of transformation into digital form, as shown in Fig. 3 and Fig. 4, respectively. Both dimensions follow a similar pattern in that the majority of the topic may be linked to certain products. The second dimension, in particular, contains terms such as nurturing (บำรุง), health (สุขภาพ), cosmetic (เครื่องสำอาง), and vitamin (วิตามิน), all of which pertain to beauty, health, and wellness-related products. The use of words such as media (สื่อ), song (เพลง), and artist (ศิลปิน)demonstrate the importance of entertainment for this issue.

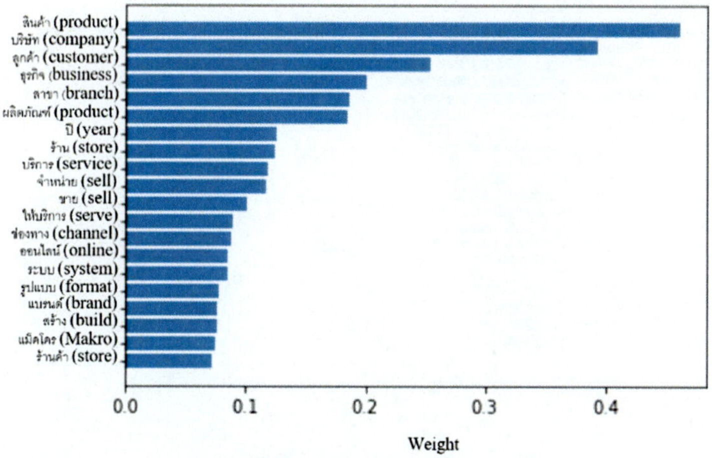

Fig. 2. Top 20 terms of latent topic 1.

The third latent topic's keyword composition is related to photography and photographic equipment. Furthermore, the use of terms such as branch (สาขา), Tesco (เทสโก้), and Lotus (โลตัส)shows that the big retail chain is a crucial component of this last dimension.

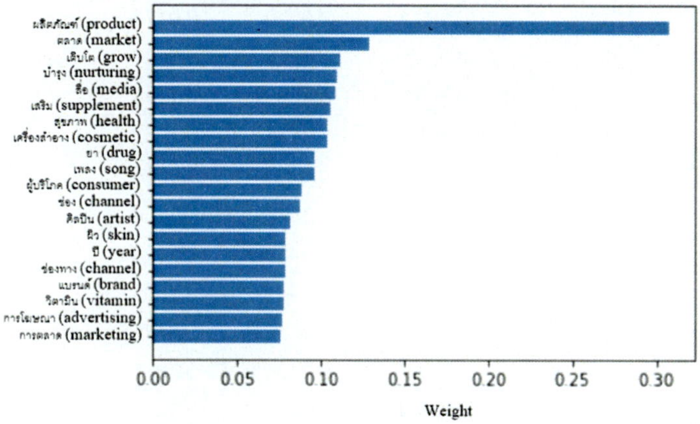

Fig. 3. Top 20 terms of latent topic 2.

We use the topic scores of the documents in consideration to categorize and assign each document into a suitable topic. Each document will be assigned to the topic with the highest score out of all the topics for which it received a score. For example, the first document's scores as shown in Table 1 suggest that it belongs to topic 1 because it received the maximum point (0.7235) for that topic.

According to our document topic matrix, the obtained topics scores suggest that 12 of the 14 firms in the sample fall into the first latent dimension. The scores also indicate that there is just one firm in each group of the second and third latent dimensions. It

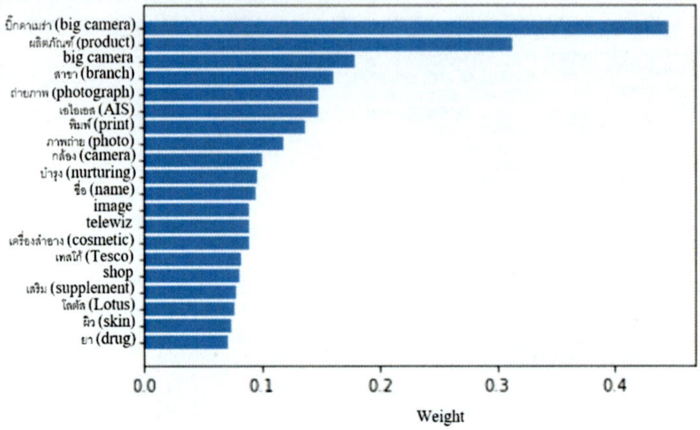

Fig. 4. Top 20 terms of latent topic 3.

Table 1. Topic scores for each document.

Document	Topic 1	Topic 2	Topic 3
1	0.7235	−0.3300	−0.0971
2	0.6441	−0.1013	−0.0176
3	0.7937	0.0620	−0.0176
4	0.6335	0.1656	−0.0824
5	0.7754	−0.1167	−0.1095
6	0.8075	−0.1175	−0.1062
7	0.6642	0.0640	−0.2513
8	0.7605	0.1881	0.1760
9	0.3720	−0.4464	0.7120
10	0.6007	−0.0437	−0.1605
11	0.5240	0.3654	0.0375
12	0.6418	0.3185	0.2031
13	0.4645	0.4982	0.3235
14	0.6663	−0.4282	0.0264

should be noted that the majority of the enterprises listed in the first topic are well-known hypermarkets or discount stores, such as Home Product Center, CP All, and Siam Makro. Table 2 shows a list of all firms in the sample and their relevant topics.

Table 2. List of firms belong to each topic.

Topic 1	Topic 2	Topic 3
RS	Mega Lifesciences	BIG Camera Corporation
Com7		
CP All		
Karmarts		
MC Group		
B-52 Capital		
Siam Makro		
Berli Jucker		
Siam Global House		
Home Product Center		
Beauty Community		
Communication & System Solution		

5 Discussion

This research focuses on how successful or rapidly growing commerce firms design their business models in response to changing business environments and competitive landscapes. It should be mentioned that, despite the fact that the firms in our sample are from the same business category on the SET, there are differences in their basic business operations. Although they all sell finished goods to clients, their products range from consumer goods to building materials and even health-related items. Furthermore, while many of the companies in the sample are well-known modern trade merchants with physical branch coverage and distribution center management as essential success factors, there is one media-based company that uses telesale as a selling channel. These components have a substantial impact on the evolution of their business model.

Another thing worth mentioning is the presence of a few key phrases that are related to the generated latent topic. As previously stated, we expect to find less frequent words when using the TF-IDF as the text feature weight calculation method because the approach gives greater values for rare terms in the document corpus. Surprisingly, we discovered that some of the important terms for each hidden topic are words that should have been common in business reports, such as products, client, or branch. We assume this is due to the algorithm's word frequency element dominating the findings, as these terms appear frequently in our corpus.

Because 12 of the 14 businesses in the sample belong to the first latent dimension, we can conclude that developing online sales channels alongside traditional business forms is the key business model transformation approach for those who want to survive and thrive in the digital era. This is especially true for well-known hypermarkets or discount stores such as Home Product Center, CP All, and Siam Makro. Because various digital-based competitors are challenging the retail business, these outcomes are in line with our expectations. These disruptors might be global businesses like Amazon or Alibaba, or regional marketplaces like Shopee and Lazada. The findings suggest that incumbents can no longer rely on their old existing competitive advantage to continue their dominance in the sector, and that the most recent obvious solution may be to migrate into the digital sphere.

Despite the fact that the key phrases collected for the second and third latent themes did not reveal any obvious business model transformation, it is observed that firms in some hidden categories benefit indirectly from digital-related developments. This is the case with Big Camera Corporation, a national retailer of digital cameras with over 200 sites. Its business solutions align with current consumer demand for shooting images and sharing them on a variety of social media channels.

Although the data utilized in this study is in Thai, the approach is applicable to texts in any languages. However, most languages have diverse structures. In addition, the cultural background linked with the language may lead to varied degrees of information disclosure. As a result, the raw inputs should be preprocessed to match the language's structure. This requirement ensures that the same technique can be utilized reliably in any target languages.

6 Conclusion

This study investigates the transformation of business models among leading corporations in Thailand's commerce sector that are publicly traded on the SET. The sample firms are screened based on their historical sales success as well as their potential for growth. Using LSA to evaluate business texts from a company's annual report, the researchers discovered that successful retailers use digital business model change in response to digital technological disruption. Even among incumbents with a strong market position, this behavior is common. As a result of the findings, moving business models toward the digital arena is a significant agenda item for all types and levels of commerce businesses.

Because this study focuses solely on the transformation of a successful or fast-growing company's business model, it is possible to conduct more research in the same issue, but this time using failing companies as a source of information. It would be interesting to see if the hidden topics extracted from lower-tier firms are similar to or different from those extracted from top-tier firms. Furthermore, the findings of such research should not only validate the findings of this study, but they may also reveal some unexpected insights.

The findings also suggest another potential future research area. The success or failure patterns of digital transformation are one factor to evaluate. According to data, the majority of digital projects do not generate the promised results. This could be owing to the process's complication, as digital transformation is described as the use of digital technologies to improve all aspects of a business, including the business model, product, process, and organizational structure [18]. Furthermore, digital transformation is characterized as a continuous process that can take a specified path over a set period of time. As a result, the procedure is multidisciplinary in character and complex. It should be mentioned that the demand for the digital transformation cookbook is not confined to the retail sector, but also includes other concerned industries such as finance and banking, tourism, real estate, and so on.

References

1. Hagel III, J., Brown, J.S.S.T.L.K.M., Samoylova, T., Lobaugh, K., Goel, N.: The retail transformation: Cultivating choice, experience, and trust. Deloitte University Press (2015). https://www2.deloitte.com/content/dam/Deloitte/za/Documents/consumer-business/ZA_RetailTrans_ConsumerBusiness_101515.pdf. Accessed 30 Dec 2021
2. Reinartz, W., Wiegand, N., Imschloss, M.: The impact of digital transformation on the retailing value chain. Int. J. Res. Mark. **36**(3), 350–366 (2019)
3. Verhoef, P.C., et al.: Digital transformation: a multidisciplinary reflection and research agenda. J. Bus. Res. **122**, 889–901 (2021)
4. Salvi, A., Vitolla, F., Rubino, M., Giakoumelou, A., Raimo, N.: Online information on digitalisation processes and its impact on firm value. J. Bus. Res. **124**, 437–444 (2021)
5. Lee, B., et al.: About relationship between business text patterns and financial performance in corporate data. J. Open Innov. Technol. Mark. Complex. **4**(1), 1–18 (2018). https://doi.org/10.1186/s40852-018-0080-9
6. Gupta, R., Gill, N.S.: Financial statement fraud detection using text mining. Editor. Preface **3**(12), 189–191 (2012)
7. Krstić, Ž., Seljan, S., Zoroja, J.: Visualization of big data text analytics in financial industry: a case study of topic extraction for Italian banks. In: Proceedings of the ENTRENOVA-ENTerprise REsearch InNOVAtion Conference, pp. 35–43 (2019)
8. Guo, L., Xu, L.: The effects of digital transformation on firm performance: evidence from China's manufacturing sector. Sustainability **13**(22), 12844 (2021)
9. Chen, W., Srinivasan, S.: Going Digital: Implications for Firm Value and Performance. Harvard Business School (2019). https://mitsloan.mit.edu/sites/default/files/inline-files/Going%20Digital.pdf, Accessed 30 Dec 2021
10. Eremina, Y., Lace, N., Bistrova, J.: Digital maturity and corporate performance: the case of the Baltic states. J. Open Innov. Technol. Mark. Complex. **5**(3), 54 (2019)
11. Kountouri, I., Manousakis, E., Tsekrekos, A.E.: Latent semantic analysis of corporate social responsibility reports (with an application to Hellenic firms). Int. J. Discl. Gov. **16**(1), 1–19 (2019). https://doi.org/10.1057/s41310-018-0053-z
12. Pröllochs, N., Feuerriegel, S.: Business analytics for strategic management: identifying and assessing corporate challenges via topic modeling. Inf. Manage. **57**(1), 103070 (2020)
13. Anandarajan, M., Hill, C., Nolan, T. (eds.): Practical Text Analytics. AADS, vol. 2. Springer, Cham (2019). https://doi.org/10.1007/978-3-319-95663-3
14. Sarkar, D.: Text Analytics with Python. Apress, New York (2016)
15. Rahmati, P., Tafti, A.R., Westland, J.C., Hidalgo, C.: When all products are digital: complexity and intangible value in the ecosystem of digitizing firms. MIS Q. **45**(3), 1025–1058 (2021)
16. Nemlioglu, I., Mallick, S.K.: Do innovation-intensive firms mitigate their valuation uncertainty during bad times? J. Econ. Behav. Organ. **177**, 913–940 (2020)

17. Camino-Mogro, S., Armijos, M., Vera-Gilces, P.: High-growth firms and international trade: evidence from Ecuador. J. Evol. Econ. 1–34 (2022)
18. Vial, G.: Understanding digital transformation: a review and a research agenda. J. Strateg. Inf. Syst. **28**(2), 118–144 (2019)

Drugtionary: Drug Pill Image Detection and Recognition Based on Deep Learning

Naphat Pornbunruang, Veerapong Tanjantuk, and Taravichet Titijaroonroj(✉)

Faculty of Information Technology, King Monkut's Institute of Technology Ladkrabang, Bangkok, Thailand
{61070044,61070213,taravichet}@it.kmitl.ac.th

Abstract. Drugtionary, which is a mobile application, is developed to support people who lack medical understanding and avoid taking the wrong drug. It consists of four main features including (i) sign-up, (ii) managing profile and medication history, (iii) viewing medication information, and (iv) managing the schedule. For viewing medication information, there are three ways to retrieve the drug information–(i) text search, (ii) chatbot, and image search. We use string search and DialogFlow for text search and chatbot, respectively, whereas deep learning technique for image detection and recognition is used to search the given drug pill image. The experimental result shows that the model generated from the CenterNet method is suitable when compared to the Faster-RCNN, RetinaNet, Yolo, and SSD on our drug pill dataset. Moreover, our application is constructed by using React and React Native technology. All data are stored in the MongoDB database.

Keywords: Drug pill image · Deep learning · Image detection · Image recognition

1 Introduction

Nowadays, medicine is very important to humans. Medicine has been used to treat ailments if an illness occurs. The patient will go to the hospital for treatment, and then the doctor will prescribe medication to them. In some cases, the patients need to reduce medical expenses [1]. They go to the pharmacy to buy medicine and take it by themselves without knowledge about the drug, such as its properties, the amount of intake, etc. The use of the wrong drug and misunderstanding are the problems. From information on pharmacists at a hospital, we found that one group of patients, most of whom were over 45 years, contacts the pharmacist for more information about the drug because the drug label is damaged or the drug is stored for a long time until forgetting its details. Thus making pharmacists' workload increase, especially, under the Covid-19 epidemic crisis, causing paramedics to have more workloads.

Recently, many researches aim to detect and recognize the drug pill from the given image. Before 2012, they applied the pattern recognition concept based on

P. Meesad et al. (Eds.): IC2IT 2022, LNNS 453, pp. 43–52, 2022.
https://doi.org/10.1007/978-3-030-99948-3_5

handcraft-feature technique to achieve it. For example, In 2010, Lee and his colleague [2] attempt to develop illicit drug identification to reduce influencing criminal activity from drug pill image. They have applied Hu moments and grid intensity for the feature extraction and L2-norm for classification. In 2011, Kim et al. [3] demonstrated the application for drug pill detection to support the elder and people who lack knowledge about medication. It employs shape classification and color matching to retrieval the drug information. After 2012, deep learning was established and it has become so popular until now, for instance, In 2018, Ou et al. [1] presented the automatic system to detect the drug pill from the given image by using a convolution neural network. In 2020, Liu and his colleague [4] aims to detect illegal, illicit, and dangerous drugs by developing a drug label identification system. It was constructed from a connectionist text proposal network (CTPN) in conjunction with Tesseract OCR Engine. We found that the researches as mentioned above aim to reduce the time-consuming and high labor-cost.

In order to reduce the paramedic's workload, especially with the pharmacist, the Drugtionary application is proposed to alleviate the above issues by collecting drug information such as properties, dosage, and precautions in the application. Our mobile application can support the user search by (i) text search, (ii) chatbot, and (iii) image search. Firstly, text search is developed by using a string matching technique. Secondly, a chatbot is produced by using the concepts of the sentiment analysis via DialogFlow [5]. Lastly, image search is constructed by using the existing deep learning techniques for image detection and recognition–Faster R-CNN [6], RetinaNet [7], SSD [8], Yolo [9], and CenterNet [10]. Additionally, our application can remind the time of taking medicines to ensure consistent medication intake. Moreover, it also stores the patient's drug use and drug allergy history that can be shown to the pharmacist or doctor immediately. For this paper, we focus on the image search based on the deep learning technique.

The remainder of this paper consists of three main sections. Firstly, the details of Drugtionary application are presented in Sect. 2. Secondly, the experimental results are appeared and discussed in Sect. 3. Lastly, the conclusion is given in Sect. 4.

2 Details of Drugtionary Application

This section aims to introduce the overall of the Drugtionary application and describe the details of each search. Therefore, this section is divided into three subsections including (i) overview of Drugtionary application, (ii) image search, and (iii) text search and chatbot as follows.

2.1 Overview of Drugtionary Application

Drugtionary application consists of four main features including (i) sign-up, (ii) managing profile and medication history, (iii) viewing medication information, and (iv) managing the schedule. The graphic user interfaces (GUI) are shown in Fig. 1. They are designed to support older people who lack medical understanding of taking the wrong drug. Our application is built on a cross-platform mobile application using React and React Native. The software architecture of the Drugtionary application is summarily displayed in Fig. 2. Moreover, MongoDB, which is a NoSQL document database, is used to store the user, schedule, and drug data. It is a flexible schema, meaning that each document does not have to have the same structure, attribute name, and data type. Lastly, DialogFlow and PyTorch are used to implement the chatbot and image search, respectively.

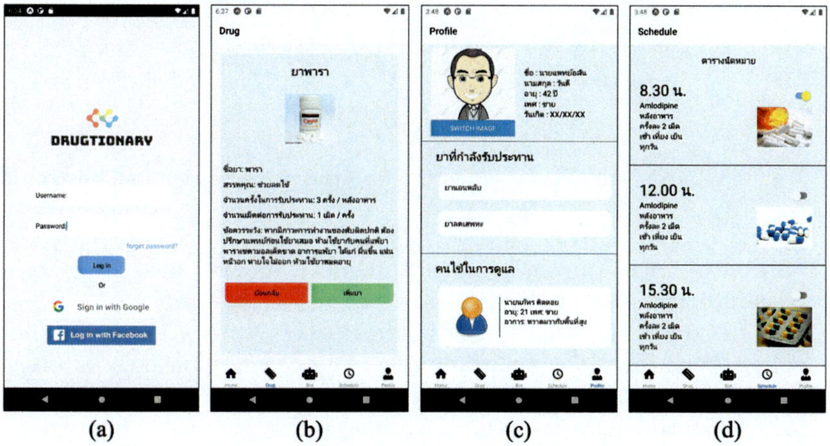

Fig. 1. Graphic user interfaces of the drugtionary application.

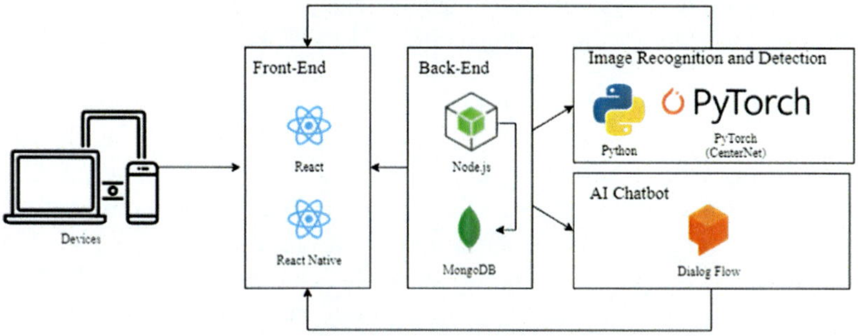

Fig. 2. An architecture of the drugtionary application.

2.2 Image Search

To search the drug information from the given image, the deep learning technique is used. Nowadays, there are many architectures such as Faster R-CNN [6], RetinaNet [7], SSD [8], Yolo [9], and CenterNet [10]. Those can be used to construct the model for detecting the wanted objects in the whole image and predicting the correct class name. Mainly, the architecture can be divided into

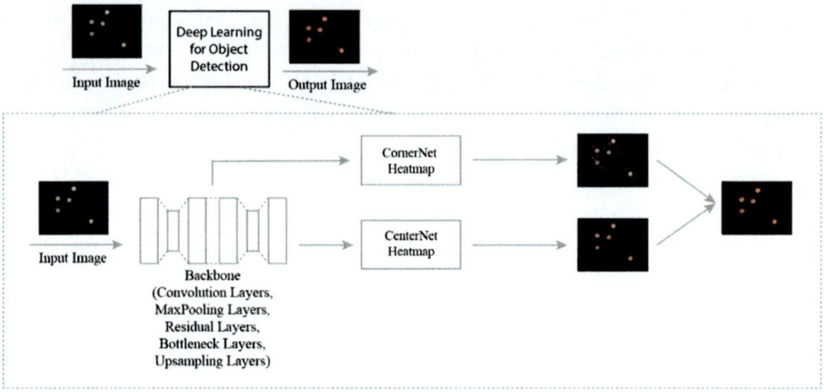

Fig. 3. A simplified CenterNet architecture for drug pill detection and recognition.

four main parts–(i) feature map extraction, (ii) bounding box detection, (iii) object classification, and (iv) bounding box regression as follows:

Firstly, to extract the rich information from the given image, the convolution and pooling layers play an important role. The convolution layer is designed based on the convolution operation with the learning filter. This layer is used to extract the feature map from the given image. The number of filters is different depending on each architecture. Furthermore, the pooling layer is useful to reduce the feature map size. It leads to reduce the computing time and produce the feature map robust to the scaling, shift, and rotational invariants.

Secondly, the bounding box can be detected from the whole image. There are many solutions to achieve it such as (i) selective search, (ii) anchor box technique with region proposal network (RPN), (iii) key point technique, and so on. The selective search is an initial solution used in deep learning for object detection and recognition as presented in [11]. It considers the characters of each pixel and merges them together. Although it is successful for detecting the bounding box, it is slow as reported in [6]. After that, the anchor box technique in conjunctions with the RPN is proposed to reduce time-consuming and increase the efficiency of the detector. However, there are two main drawbacks as described in [12]. It produces a large set of anchor boxes for training the accurate model. The anchor boxes have many parameters, e.g. amount of box, box size, and aspect ratios that users have to define manually. Finally, the key point technique aims to overcome the drawbacks above. A pair of corner key points including top-left and bottom-right corners are used. They are extracted from the feature map before grouping them together. All solutions as mentioned above can produce the bounding box from the whole image.

After that, the feature map and the bounding box obtained from the first and second parts are used to predict the correct class label of its bounding box by using existing classification methods such as support vector machine (SVM) or neural network (NN). Many architectures almost require the transforming step from the feature map to the feature vector before feeding it to classify. Moreover, if the architecture cannot support a variety of the bounding box sizes, the feature map must be resized before.

Lastly, the bounding box obtained from the second part is still inaccurate. Therefore, to enrich the bounding box, the bounding box regression technique is required. It is achieved by using linear regression and can produce the bounding box more accurate. Based on our experiment in Sect. 3, The CenterNet is appropriate for drug pill detection and recognition and its simplified architecture is displayed in Fig. 3.

2.3 Text Search and Chatbot

They are an alternative solution to support the user who wants to get the drug information by typing text. For text search, the user can type the drug name or its detail in the text input. After that, the application will show the drug information corresponding to your input as shown in Fig 4. It is a real-time search, meaning that the result is changed whenever you type.

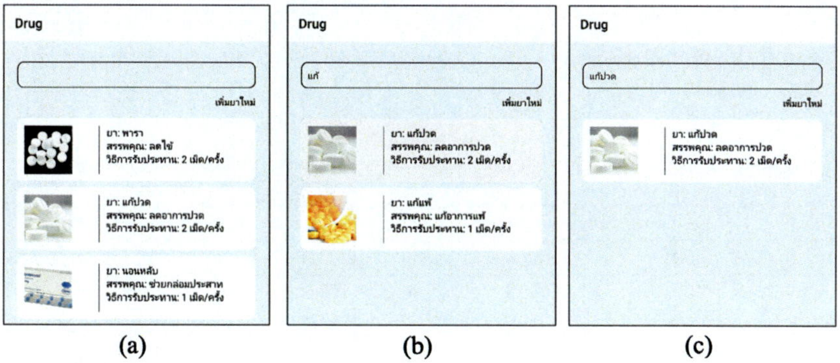

Fig. 4. An example of searching drug information by typing text.

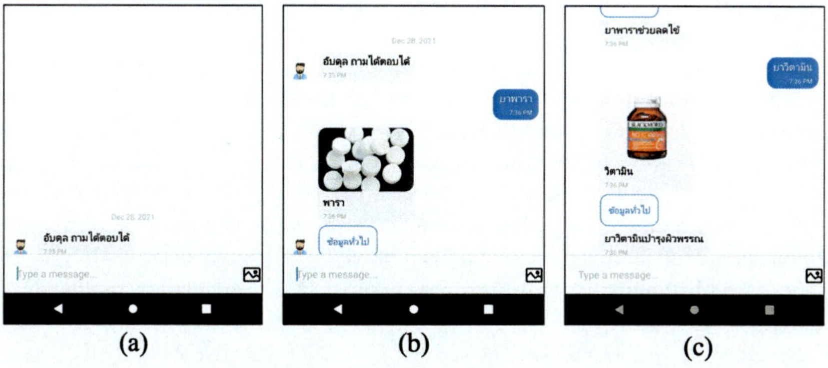

Fig. 5. An example of searching drug information by using chatbot.

In our application, the chatbot is developed by using DialogFlow [5]. It is a widely used and well-known tool to build the chatbot easily. The DialogFlow is constructed based on the sentiment analysis technique in the natural language processing field. To understand the coming message from the user, the DialogFlow is trained by drug name and its detail. While the response message is set by the drug information including name, properties, and link for viewing its detail as shown in Fig. 5.

3 Experiment

This section aims to show the performance of the existing deep learning techniques and select the best one for the image search in our application. Experimental setup and results are given below.

3.1 Experimental Setup

In order to compare the performance of the existing deep learning techniques for drug pill image detection and recognition, Faster-RCNN [6], RetinaNet [7], SSD [8], Yolo [9], and CenterNet [10] are selected here. Their parameters such as learning method, learning rate, epoc, weight decay, and loss function are summarized in Table 1.

Table 1. The configuration of each method used in this experiment.

Method	Learning rate	Weight decay	Epoch	Learning method	Loss function
Faster R-CNN	0.001	0.0005	48	SGD	Cross entropy loss
RetinaNet	0.001	0.0010	36	Adam	Focal loss
SSD	0.001	0.0005	36	SGD	Cross entropy loss
Yolo	0.001	0.0005	36	SGD	Focal loss
CenterNet	0.001	-	168	SGD	Focal loss

*Stochastic gradient descent (SGD) is a well-known learning method.

This experiment is set up by using a drug pill dataset containing 690 images with resolutions 4096 $\times$ 3024 pixels from 17 frequently used drugs including Ibuprofen, Paracetamol, Amoxicillin, Domperidone, Diclofenac, VitaminB-Complex, Cetirizine, Chlorphenamine, Dicloxacillin, Losartan, Glipizide, Amlodipine, Simvastatin, Aspirin, Metformin, Enalapril, and Hydrochlorothiazide as recommended from pharmacists at a hospital. All images are captured by ourselves using smartphone in JPEG format. This dataset is divided into two sets–(i) 639 images for training image and (ii) 51 images for testing image, randomly.

The precision (p), recall (r), f1-score (f), and false discovery (fd) are used to evaluate the performance for the image detection and the accuracy rate (a) is used for image classification. The higher values are better while lower values are worse for p, r, f, and a whereas higher values are worse while lower values are better for fd. To easily understand, we have converted the range of the precision, recall, f1-score, and false discovery to 0–100%.

3.2 Results and Discussions

Based on the experimental result in Table 2, it shows that the accuracy rate of the Yolo and SSD methods had the highest accuracy at 100%. Moreover, the SSD method still got the highest precision value at 100%. In terms of recall, the CenterNet had the highest recall value at 73.37% while the other methods received recall values less than 50%. The CenterNet method got the highest f1-score value which denotes the overall performance for the image detection. Finally, the false discovery value indicates the error per predicted number. The lowest value is 0% obtained from the SSD method.

Table 2. A comparison between the performance of Faster-RCNN, RetinaNet, SSD, Yolo, and CenterNet.

Methods	Image detection				Image recognition
	Precision	Recall	F1-score	False discovery	Accuracy rate
Faster R-CNN	98.70	40.86	57.79	1.30	98.70
RetinaNet	96.83	32.80	49.00	3.17	96.83
SSD	**100.00**	29.95	46.09	**0.00**	**100.00**
Yolo	97.67	22.58	36.68	2.33	**100.00**
CenterNet	97.12	**73.37**	**83.59**	2.87	97.83

The example results of the Faster-RCNN, RetinaNet, SSD, Yolo, and CenterNet methods are shown in Tables 3 and 4. The detected drug pills are presented in red boundary boxes. The CenterNet can almost detect drug pills in the given image except for the case of two pills that are close to each other, since it cannot locate a clear bounding box. However, the detected number of the Faster-RCNN, RetinaNet, Yolo, and SSD is lower than that of the CenterNet because they detect the bounding box by using the anchor box techniques. This technique has fixed three box ratios (width per height) and three box sizes with the RPN model. It might be not appropriate for the large image because our training dataset is quite small and its image size is large. Although a pre-train model from the COCO dataset, which is widely used, is employed, it cannot enrich the performance of the Faster-RCNN, RetinaNet, Yolo, and SSD in this case.

To choose the appropriate deep learning method to serve the image search in our application, the performance should be not considered only one measure. We need a model that is balanced from all measures above. Therefore, the model constructed from the CenterNet is suitable for this case. It has a relatively high accuracy rate, precision, recall, f1-score and still low false discovery when compared to the Faster-RCNN, RetinaNet, Yolo, and SSD.

Table 3. The example of resulted images from each method on Cetirizine, Chlorphenamine, Dicloxacilline, Ibuorofen, Paracetamol, Amoxicillin, Domperidone, Diclofenac, Vitamin B Complex, and Losartan.

Drug Name	Faster R-CNN	RetinaNet	SSD	Yolo	CenterNet
Cetirizine	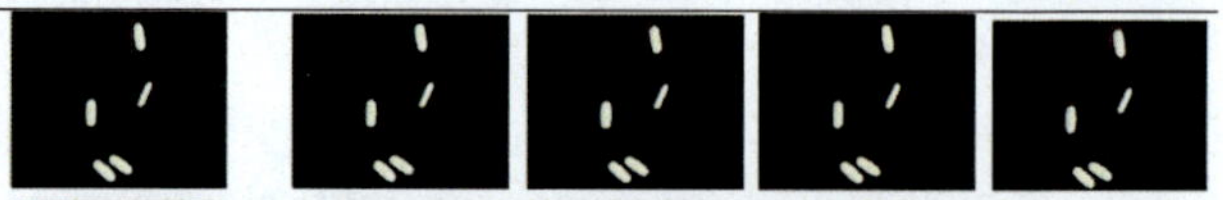				
Chlorphenamine	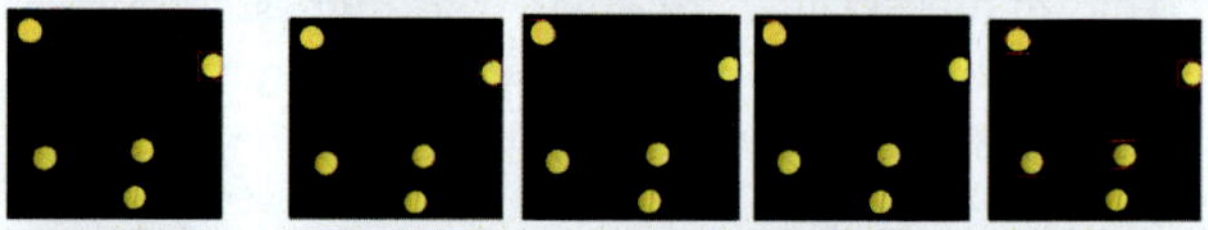				
Dicloxacilline	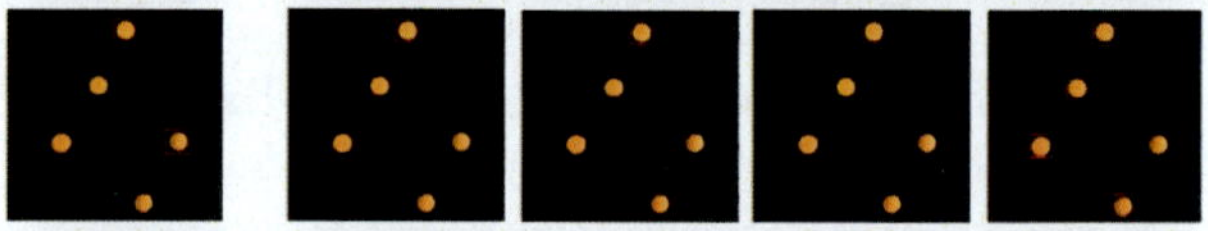				
Ibuorofen	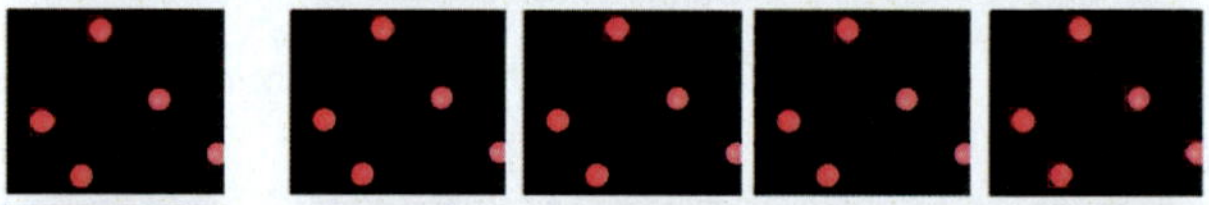				
Paracetamol	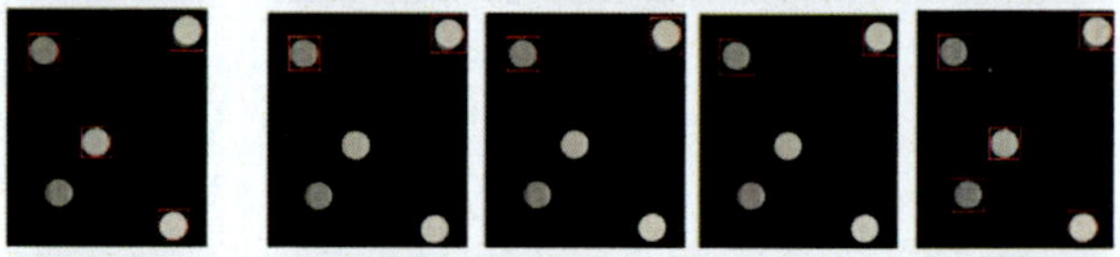				
Amoxicillin	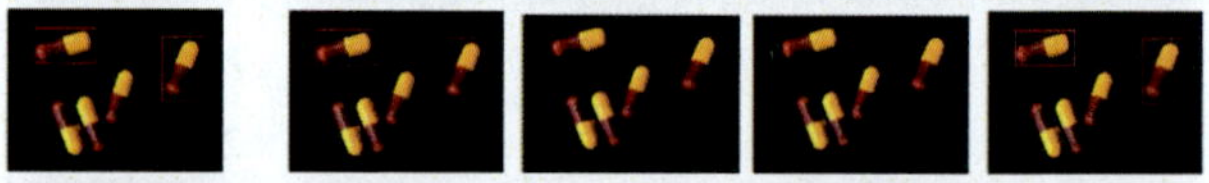				
Domperidone	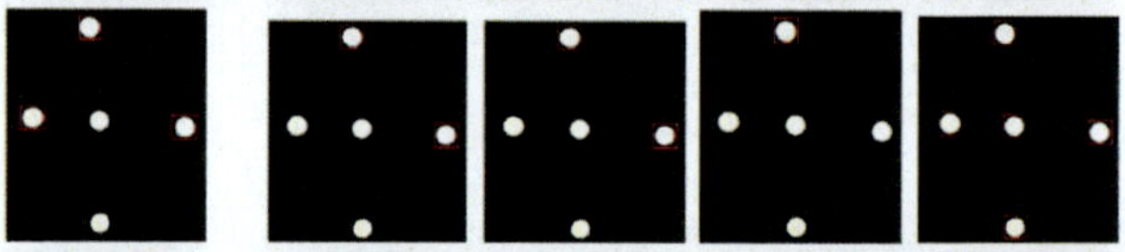				
Diclofenac	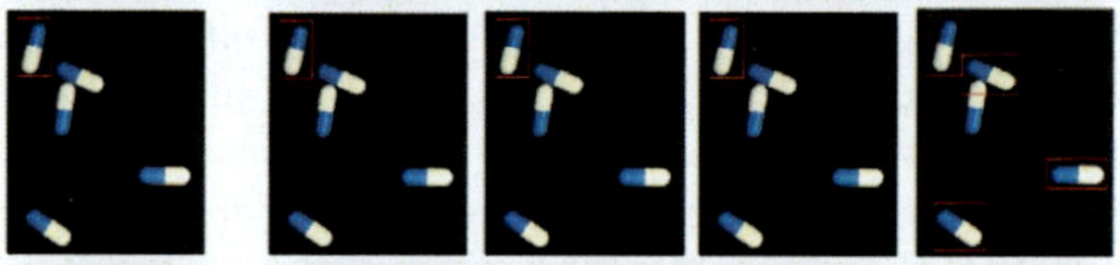				
Vitamin B Complex	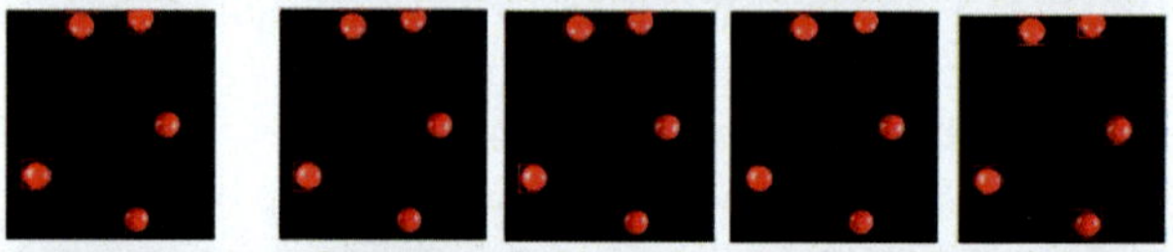				
Losartan	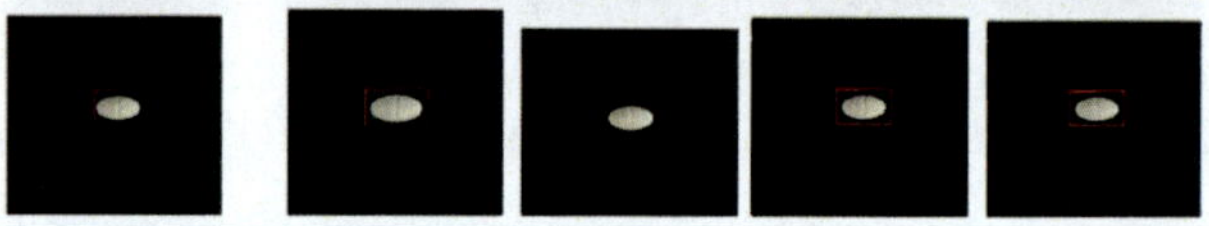				

Table 4. The example of resulted images from each method on Glipizide, Amlodipine, Simvastatin, Aspirin, Metformin, Enalapril, and Hydrochlorothiazide.

Drug Name	Faster R-CNN	RetinaNet	SSD	Yolo	CenterNet
Glipizide	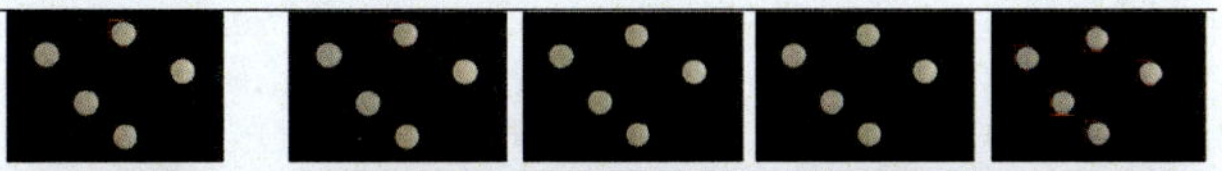				
Amlodipine	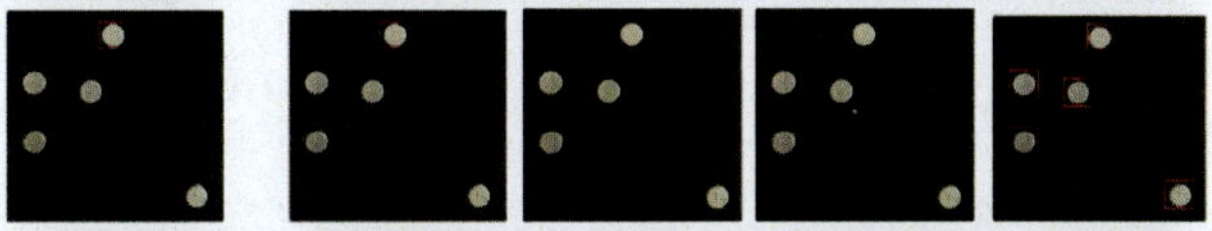				
Simvastatin	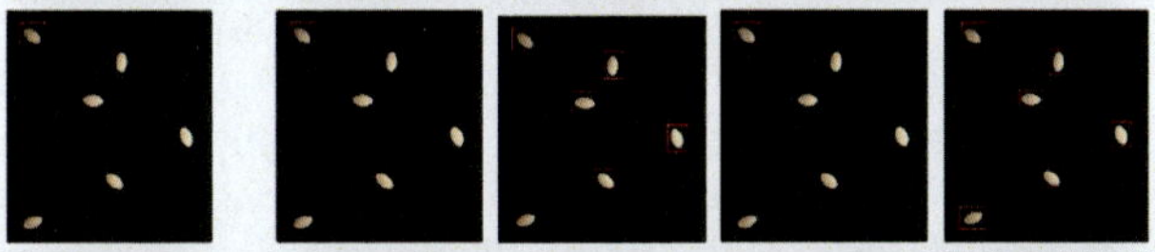				
Aspirin					
Metformin	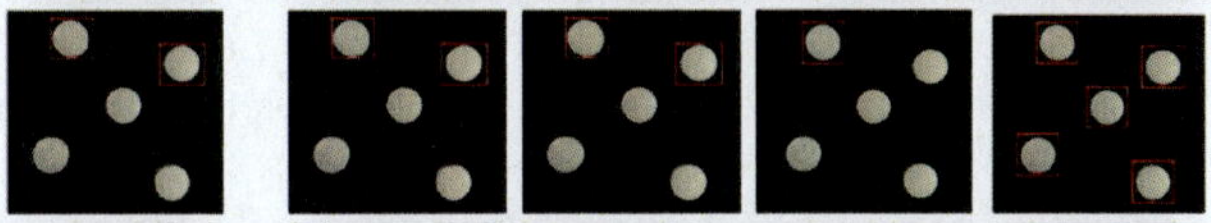				
Enalapril	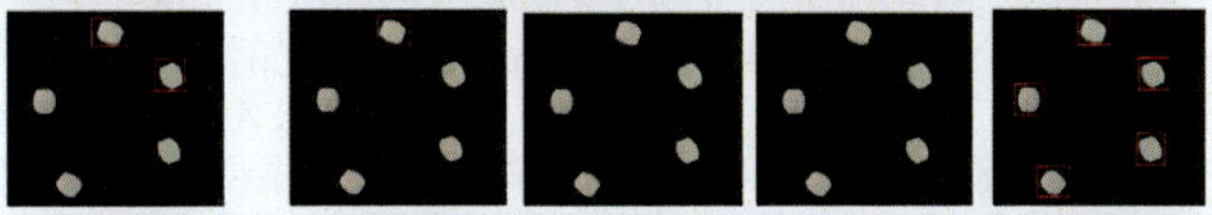				
Hydrochlorothiazide	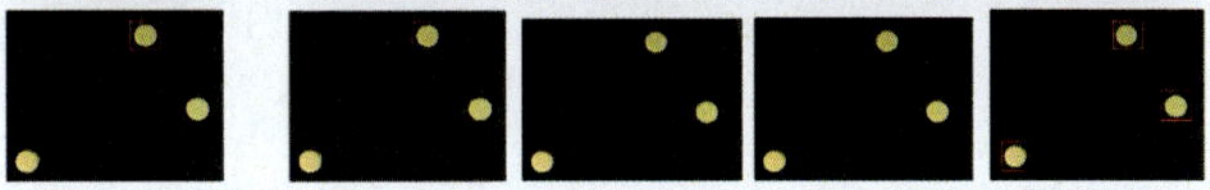				

4 Conclusion

In this paper, the Drugtionary, which is a mobile cross-platform application, is proposed. It can be used to support the elder and people who lack knowledge about medication. Users can retrieve the drug information in three ways including text search, chatbot, and image search. String search and DialogFlow techniques are selected to develop the text search and chatbot, respectively. However, this paper focuses on the image search. We aim to find the appropriate deep learning technique for the drug pill image detection and recognition. Based on the experimental result, the CenterNet is suitable for our application when measured it in term of the precision, recall, and false discovery for the image detection and the accuracy rate for the image recognition. Moreover, our application can remind the time of taking medicines and store the patient's drug use. This may help the doctor to treat the patient.

References

1. Ou, Y.Y., Tsai, A.C., Wang, J.F., Lin, J.: Automatic drug pills detection based on convolution neural network. In: 2018 International Conference on Orange Technologies (ICOT), pp. 1–4. IEEE, Nusa Dua, BALI, Indonesia (2018)
2. Lee, Y.B., Park, U., Jain, A.K.: PILL-ID: matching and retrieval of drug pill imprint images. In: 2010 20th International Conference on Pattern Recognition, pp. 2632–2635. IEEE, Istanbul, Turkey (2010)
3. Kim, D., Chun, J.: Drug image retrieval by shape and color similarity of the medication. In: 2011 First ACIS/JNU International Conference on Computers. Networks, Systems and Industrial Engineering, pp. 387–390. IEEE, Jeju, Korea (South) (2011)
4. Liu, X., Meehan, J., Tong, W., Wu, L., Xu, X., Xu, J.: DLI-IT: a deep learning approach to drug label identification through image and text embedding. BMC Med. Inf. Decis. Mak. **20**, 68 (2020)
5. Google Dialog Flow. Available at January 3, 2022. https://dialogflow.cloud.google.com/. Accessed 03 Jan 2022
6. Ren, S., He, K., Girshick, R., Sun, J.: Faster R-CNN: towards real-time object detection with region proposal networks. In: Cortes, C., Lawrence, N., Lee, D., Sugiyama, M., and Garnett, R. (eds.) Advances in Neural Information Processing Systems. Curran Associates, Inc. (2015)
7. Lin, T.Y., Goyal, P., Girshick, R., He, K., Dollar, P.: Focal loss for dense object detection. IEEE Trans. Pattern Anal. Mach. Intell. **42**(2), 318–327 (2020)
8. Liu, W., Anguelov, D., Erhan, D., Szegedy, C., Reed, S., Fu, C.-Y., Berg, A.C.: SSD: single shot MultiBox detector. In: Leibe, B., Matas, J., Sebe, N., Welling, M. (eds.) ECCV 2016. LNCS, vol. 9905, pp. 21–37. Springer, Cham (2016). https://doi.org/10.1007/978-3-319-46448-0_2
9. Redmon, J., Divvala, S., Girshick, R., Farhadi, A.: You only look once: unified, real-time object detection. In: 2016 IEEE Conference on Computer Vision and Pattern Recognition (CVPR). pp. 779–788. IEEE, Las Vegas, NV, USA (2016)
10. Duan, K., Bai, S., Xie, L., Qi, H., Huang, Q., Tian, Q.: CenterNet: keypoint triplets for object detection. In: 2019 IEEE/CVF International Conference on Computer Vision (ICCV), pp. 6568–6577. IEEE, Seoul, Korea (South) (2019)
11. Girshick, R., Donahue, J., Darrell, T., Malik, J.: Rich feature hierarchies for accurate object detection and semantic segmentation. arXiv:1311.2524 [cs]. (2014)
12. Law, H., Deng, J.: CornerNet: detecting objects as paired keypoints. Int. J. Comput. Vis. **128**(3), 642–656 (2019). https://doi.org/10.1007/s11263-019-01204-1

Integrating Fishbone Diagram from Descriptive and Predictive Data Mining for Describing the Relation Between Cardiovascular Diseases and Related Items

Sawinee Sangsuriyun, Peera Liewlom(✉), Surasak Tangsakul, and Sasithon Suchaiya

Faculty of Science and Engineering, Kasetsart University, Sakon Nakhon 47000, Thailand
{savinee.sa,peera.li,surasak.tang,ssrithon.s}@ku.th

Abstract. This research is aimed to discover the relationship between cardiovascular disease and related factors through a Knowledge Discovery in Databases (KDD) by looping a variety of data mining techniques with both descriptive and predictive data mining. These techniques are the k-Mean Clustering, the Association Rule Discovery, the Association Rule Tree Reasoning, and the C4.5 Decision Tree Classification. The outcome is a fishbone diagram integrating the results from these data mining techniques. It consisted of 5 main topics, 2 main topics with ordered factors, and 16 environmental factors discovered from 918 cases of the Heart Failure Prediction Dataset.

Keywords: Fishbone diagram · Data mining · Association rule · Clustering · Decision Tree · Cardiovascular diseases

1 Introduction

Cardiovascular diseases are the main cause of death worldwide [1], with 17.9 million deaths per year, and the mortality rate is continually rising. Thailand has 58,681 annual fatalities, or an average of seven individuals every day [2]. Heart disease has a wide range of reasons, and more research studies on the condition are being conducted every year. As of December 5, 2021, a search of the Scopus database reveals more than 40,000 studies every year in this field, indicating that there will be an increase in cardiovascular knowledge to minimize mortality due to cardiovascular disease.

The presence of a huge number of patients all over the world, as well as a vast amount of data or datasets, allows researchers to understand more about the condition. This makes knowledge discovery from huge datasets [3] a technique that can aid in our understanding of cardiovascular disease. Data mining [4], which substantially helps the rapid discovery of important patterns in datasets,

P. Meesad et al. (Eds.): IC2IT 2022, LNNS 453, pp. 53–67, 2022.
https://doi.org/10.1007/978-3-030-99948-3_6

is one of the important techniques in the knowledge discovery process. It serves as a crucial basis for further knowledge development.

The data mining techniques utilized in this research were used to find interesting patterns hidden in the new cardiovascular dataset from kaggle.com, which was published in September 2021 [5]. The focus is on describing or discovering specific or interesting characteristics or relationships between patient qualifications and heart disease.

This study uses descriptive data mining and uses three techniques: 1) k-Mean [4]; 2) the Apriori algorithm for association rule discovery to find huge itemsets [6]; and 3) association rule tree reasoning [7], a predictive data mining technique capable of explaining the relationship between patient features and cardiovascular diseases by adjusting parameters for descriptive purposes, which is the C4.5 Decision Tree Technique [8]. All of the approaches utilized in this study are the outcome of each technique's patterns to improve knowledge of the others. It is not a comparison to determine which model is the best. A Fishbone Diagram [9] was created as a result of the knowledge enhancement. This diagram is often used to reason about the variables involved in the topic of interest. The study methodology describes the way to expand understanding amongst various techniques.

The objectives of this paper is to integrate a number of techniques to enhance the knowledge gained from the findings of both descriptive and predictive data mining integrated into a fishbone diagram to characterize the relationship between patient with heart disease and the property values are related factors. The following sections detail the research methodology, research results, discussion, and conclusions.

2 Related Work

The study used multiple data mining techniques on the same dataset, which is the Heart Failure Prediction Dataset [5]. This dataset was published on kaggle.com in September 2021. It includes patients from five clinics in the United States and Europe. There were 918 cases in total, with 11 attributes were associated with heart disease risk factors and one class indicating whether each patient's clinical results were heart diseased or not.

For the approach to integrating knowledge from various data mining techniques. It is start by explaining the details and strengths of each technique that link with other techniques. Then describe the way for creating a fishbone diagram that was obtained by integrating knowledge from the results obtained from each technique as follows:

2.1 k-Mean Clustering

Clustering is one of the main techniques in data mining by searching for the underlying subsets in a dataset. Then describe the characteristics of each discovered subgroup by k-Mean clustering [4]. Using the attributes of each object or item in the data set, it computes the distance from the center of group k(-th). The data will be treated as a member of the nearest group, which means the shortest distance from the data to its center. The center of a group is defined by the average distance between the members of that group.

In this research, we are using the WEKA [10] data mining application for k-Mean clustering by defining k as 2 groups to supervise datasets as 2 classes and using the Euclidian distance as a distance function.

The outstanding result to be added to other techniques is to expose the characteristic features of each subgroup by choosing the obvious features that are the main characteristics of heart disease patients in this dataset.

2.2 Apriori Algorithm

The association rule discovering is another data mining technique by which rules consist of antecedent and consequent in the form of $\{Antecedent\} \rightarrow \{Consequent\}$ which means that when we find some group of antecedent attributes, we also find consequent attributes often.

According to the Apriori algorithm's association rule provided by [6], two key component values are support and confidence. Any rule or some itemset has a support value greater than the minimum support threshold, assuming the itemset rule is frequently occurring and interesting. On the other hand, if any rule has a confidence value greater than the minimum threshold, we assume the rule is stronger in associating Antecedent to Consequent and is therefore more interesting.

In this research, the WEKA application was able to discover association rules by defining minimum support as 0.1 (or not less than 10% of the dataset) and minimum confidence as 0.9 (from 1 is at most) to prevent the tremendous number of discovered rules when we want to combine with other techniques. Furthermore, we define the format of the Consequent rule as the heart disease patient class only, but the Antecedent rule is related to the heart disease patient features factor. The Consequent rule is class only, so called Class Association Rules or CARs [11].

To be used with the WEKA application, the researcher team modified some attributes of the heart disease patient dataset, especially the continuous decimal number record, which cannot be counted as a group and must be discretized as a nominal number. This data preparation is done by WEKA before the association rule, which is discovered by Consequent is heart disease only with the Apriori algorithm.

According to CARs, which expose the association rules of heart disease patients only, we can propose the fishbone diagram of itemsets or features that are associated with heart disease patients [12]. The category of fishbone diagram is all items or features that are found in all association rules. The fish head closer category has the most items occurring in all association rules, whereas the fish tail closer category has the least items occurring. The fish branch is a sub-feature that belongs to the main feature that shares the same circumstance. The items in this fishbone diagram will be filtered or grouped in the fishbone integration process later.

The fishbone association rule, where the fish head is a heart disease patient, can be merged with other fishbone diagram visualization techniques so that the fish head is a heart disease patient equivalently, as if they came from the same dataset.

2.3 Association Rule Tree Reasoning

The novel association rule tree reasoning technique was proposed by [7] to describe the sequential correlations of primary and secondary factors related to interests. In this work, the sequential correlations of these factors are used for patients with cardiovascular diseases to produce an association rule tree reasoning without specifying the lowest support and sentiment values. However, using the principle of secondary factors that complement the primary factor must promote the relationship with heart patients to be more prominent. The secondary factors usually cause a lower support value for the correlation rule. However, suppose the increasing rate of the confidence value is higher than the decreasing rate of the support value. In that case, the secondary factors can prominently promote the relationship between cardiovascular disease patients and primary factors. This relationship can be plotted as an association rule tree.

The discovery of the association rule tree was performed according to the algorithm specified in [7]. The root node of the association rule tree is defined as a cardiovascular disease patient that allows the tree to be displayed in the shape of the fishbone diagram by specifying the root node as the fish head. The primary factor category is the first factor that connects the root node as the middle axis of the fishbone, and then the next nodes can be the topics and the sub-topics, respectively.

2.4 C4.5 Decision Trees

The C4.5 Decision Trees [8] are a popular technique used in predictive data mining. It creates a prediction model in a tree-style where each node is a property in the dataset. Each edge of the tree represents an optional property value within that property node in the model. A root node is the best property node used to predict at the top of the tree structure, and the leaf node is at the bottom end of the tree structure that stands for the class value to predict. The selection of properties from root to leaf node is the most class-related property selection used to create a tree. Thus, the resulting tree model can describe the relationship between property nodes and class nodes that are, respectively, primary and secondary factors in each graph path.

This research implemented the C4.5 decision trees on the WEKA to obtain the tree model that illustrates the relationship between property nodes and class nodes as the main objective. In this regard, each leaf node was selected to display at least ten items per object data for the appropriate size of the tree. However, this parameter setting is not the best parameter for the best model. It is only an ideal parameter to convert fishbone diagrams for integrating them with other techniques. The conversion of decision trees to fishbone diagrams, used in this research, is a basic method. The fish head is defined as the cardiovascular disease patient class. Then, the tree paths are chosen from the root node to the leaf, with only the cardiovascular diseases patient class combined into the same fishbone diagram.

3 The Proposed Integrating Fishbone Diagram

The knowledge discovery process from databases is an iterative process that combines the results from each round of data mining to summarize the discovery of knowledge [4]. This research only yielded interesting patterns from different data mining techniques. The lack of verification with the current knowledge of cardiovascular diseases is a future consideration.

The integrating technique with the iteration process is still a significant step in knowledge discovery. The technique chosen in this research is characterized by a graph transformation with the destination node (root or leaf node) representing what we are interested in (i.e., a patient with cardiovascular diseases). However, these graphs or trees cannot be directly integrated due to the differences in their structures, characteristics, and operating environments, whereas their similarities are that the sequence appears in the association rule and appears as graphs in different shapes.

A fishbone diagram [9] is a technique that presents the relationship between the problem of interest and the relevant factors in an association sequence. The method focuses only on the sequence of primary and secondary factors. Therefore, if the results of any data mining are sequentially correlated with interests, they can be converted to an association sequence in a fishbone diagram without the problems of the differences in node and edge characteristics of those data mining techniques.

Therefore, fishbone diagrams can be used as a central axis (or core) for integrating the results of an association sequence in graphs obtained by different data mining techniques in each iteration with the same dataset. In this regard, the Apriori algorithm [6], association rule tree reasoning [7], and C4. 5 decision trees [8] are techniques that can be converted into a fishbone diagram (see details in Subsects. 2.2 to 2.4). The clustering result in Subsect. 2.1 is not a fishbone diagram. However, subgroup characteristics associated with cardiovascular disease patients can be determined as items or details of the fishbone diagrams obtained by other methods in this study.

In this research, we proposed the guidelines for integrating fishbone diagrams as the following steps:

Step 1. For the fishbone diagram obtained by the Apriori algorithm, the clustering of an item or property value under various risk factors has an unclear association between primary and secondary factors. Therefore, if conflicting items are found within this group in which a distinct association sequence cannot be established, then those items can be filtered out.

Step 2. Filter out the similar items or property values in every tip of the fishbone diagram derived from association rules (remaining only the parts where the items are the primary factor). Those items are distinctive features relative to the entire data set, which makes it useless to describe only the relationship to the item of interest.

Step 3. Filter out the similar items or attributes between cardiovascular diseases and non-cardiovascular diseases clusters derived from the k-Mean clustering method from all tips of the fishbone diagram obtained by association rules. Due to those items are not explicit properties of a particular subgroup but belonging to all groups.

Step 4. For the fishbone diagram obtained by association rules and filtered by the previous three steps, if the members of some primary factor are not found, and only one primary factor is unable to have a trust association rule of interest, then those factors can be filtered out. It means that the item is less significant than the others in the association's rules.

Step 5. Secondary factors in the fishbone diagrams obtained by association rules are defined by special symbols when combined with other fishbone diagrams that are only the environment in the primary factors. It differs from the secondary factors obtained by the fishbone diagram in the other two techniques with association sequence in the primary-secondary factor, respectively. Therefore, they have to illustrate different fishbone characteristics.

Step 6. The primary factor obtained by the association rule tree reasoning techniques is designated as a special symbol to show that those factors promote cardiovascular diseases. It is more significant than the primary factors from the association's rules.

Step 7. Explore the branching or herringbone elements obtained by the C4.5 decision trees. If any inconsistent elements are found when compared to the subgroup property values obtained by the k-Mean clustering, then these elements and their related elements will be removed. This includes all sub-elements that cannot represent their classes. For example, if the cholesterol of the cardiovascular disease patients is 182.63 while the cholesterol of the non-cardiovascular disease patients is 217.97, but the C4.5 branching at $<=0$ and >0, it is an unreasonable element. In this case, the entire element from the root to these elements will be removed.

Step 8. The primary factor obtained by the C4.5 decision tree technique is also defined as a special symbol. This definition shows that the secondary factors related to this primary factor and the class of interest are found.

Step 9. Combine each fishbone diagram from all techniques into a new diagram. If the graph paths in the fishbone are duplicates, then they collapse together. A goal is to show all the different symbols of all the items based on a different technique.

Step 10. Explore the fishbone diagram from the head to the tail. There is redundancy if any primary factors related to the fishbone structure (including the primary factors) are found (all similar members in that graph structure). Therefore, eliminate the part near the tail and leave only one primary factor.

Since they have similar compositions, but the meaning of the primary factor is less than the part near the tail.

Step 11. Explore the primary factor section in the fishbone diagram. If any primary items have the same property, promote it to the primary factor instead. For the old primary factor, demote it to secondary factors in that group, such as $gender = male$ and $gender = female$; gender can also be their primary and secondary factors.

Step 12. Summary of distinctive patterns that appear in the fishbone diagram obtained by integrating all techniques into categories or items/features grouped that make them easier to understand and implement.

The fishbone diagram proposed in this paper was built using XMind 8, a free downloadable version at xmind.net.

4 Experiments and Results

The Heart Failure Prediction Dataset [5] included data on 918 clinical cases from five American and European clinics. It consisted of 11 attributes or features recorded as heart disease risk factors, including Age, Sex, ChestPainType, RestingBP, Cholesterol, FastingBS, RestingECG, MaxHR, ExerciseAngina, Oldpeak, and ST Slope. As a classifier, Heart Disease was used. Thus, the preliminary data set consisted of 508 heart disease patients and 410 non-heart disease patients, 725 males and 193 females, with a mean age of 53.5 years.

From the results of the four data mining techniques used in this study, the strengths and weaknesses of each technique can be described before presenting the results of the integration of the strengths of the 4 techniques in the form of an integrated fishbone diagram as follows:

4.1 k-Means Clustering

By using the k-Means clustering technique at $k = 2$ and the Euclidean distance formula to process the heart disease dataset with the WEKA software, the results are shown in Table 1.

From Table 1, the results of clustering show that cluster 0 is a good representative of the heart disease group, and cluster 1 is a good representative of the non-heart disease group. However, most of the attribute values of the two groups are distinctly different. Any element used to integrate fishbone diagrams derived from each technique should not conflict with these values, including any similar attributes of subsets (data in the table highlighted in yellow). This indicates that such elements are not suitable for differentiating between the subsets. They should not be used for the integration of the fishbone diagram, unless a hierarchical correlation is found between primary and secondary factors associated with heart disease, where such factors are maintained in the integration of the fishbone diagram.

The advantage of clustering is that if the clustering result is close to that of the classification, then the mean of each attribute is a good representative of the

Table 1. Representative attributes of each subset of the heart disease dataset.

Attributes	Whole dataset	Cluster 0	Cluster 1
Age	53.51	56.32	50.19
Sex	M	M	M
ChestPainType	ASY	ASY	ATA
RestingBP	132.40	134.91	129.41
Cholesterol	198.80	182.63	217.97
FastingBS	No	No	No
RestingECG	Normal	Normal	Normal
MaxHR	136.81	125.08	150.71
ExerciseAngina	N	Y	N
Oldpeak	0.89	1.31	0.39
ST_Slope	Flat	Flat	Up
HeartDisease = Yes	508	430	78
HeartDisease = No	410	68	342

subset and can be used to determine the reasonableness of the results in other techniques.

The disadvantage of the k-Mean clustering is that it offers little detail about the relationships in relation to what we are interested in, requiring other techniques to elaborate on the more complex relationships between heart disease and other factors.

4.2 Fishbone Diagram Derived from Apriori Algorithm

When the heart disease dataset was customized for processing in WEKA software, only CARs of the heart disease class are required to be processed. At a minimum support level of 0.1 and minimum confidence level of 0.9, 264 association rules were obtained on the right of Consequent of Class HeartDisease = Yes and can be converted into a fishbone diagram as shown in Fig. 1.

From Fig. 1, the obtained fishbone diagram elements show high redundancy of the group of items on each branch. These duplicate elements were screened out in Step 1 to Step 4 of the integration of fishbone diagram.

The advantage of the fishbone diagram derived from the association rules is that there are many details available.

The disadvantage of this technique is that many fishbone diagrams have high redundancy in relation to what we are interested in, that is the heart disease class at "environment factors related to our interests" level. It does not imply a direct relationship between those items and our interests. Also, if the minimum support and minimum confidence levels are lowered, there will be a huge number of association rules. It is difficult to apply the results that are too detailed.

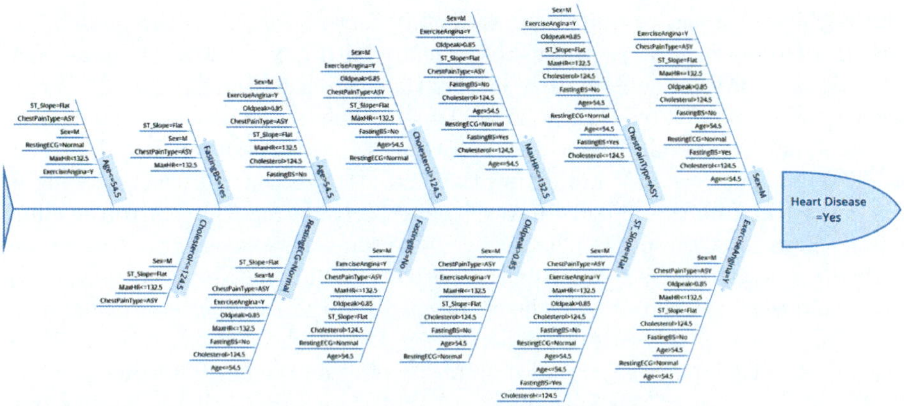

Fig. 1. A fishbone diagram derived from the association rules.

4.3 Fishbone Diagram Derived from Association Rule Tree Reasoning

When this data-mining technique was used with the same customized heat disease dataset to discover the association rules with WEKA software according to the original algorithm of this technique [7]), the result is shown in Fig. 2.

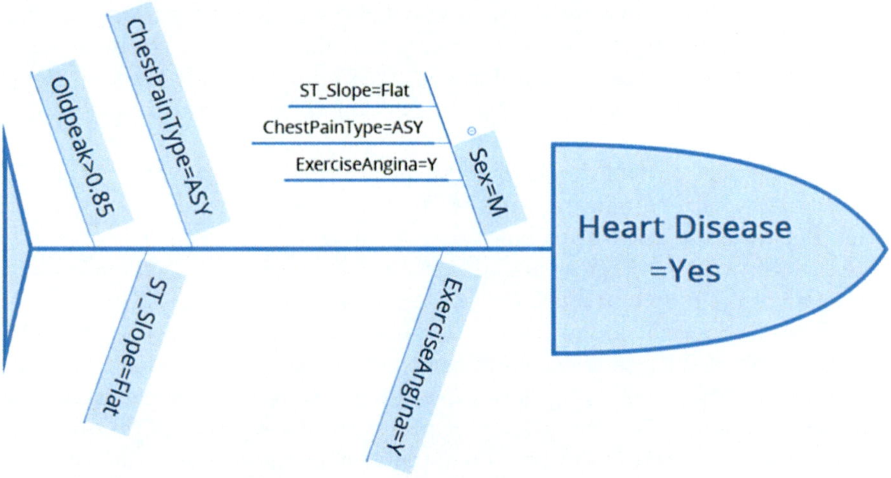

Fig. 2. Fishbone diagram derived from association rule tree reasoning.

From Fig. 2, the fishbone chart shows only the elements that promoted a stronger association with heart disease than those found with the normal association rules. There was a total of five elements. There was one factor in which the relationships between primary and secondary factors increased the association with heart disease, that is, the fishbone chart with Sex = M as the main factor. The secondary or contributory factors that increased the likelihood of heart disease are ST_Slope = Flat, ChestPainType = ASY and ExcerciseAngina = Y.

Based on these results, what was interesting was the asymptomatic chest pain or ChestPainType = ASY indicating that most of the heart disease patients in this dataset had no chest pain. However, when considering the visit to the clinic it was shown that the patient had other symptoms that suggest that he or she might have heart disease. Most of the patients in this dataset were physically active and was associated with heart disease symptoms or ExerciseAngina = Yes. The results of the pulse signal measurement of these patients showed low blood flow (ST_Slope = Flat) and high post-exercise pressure (Oldpeak > 0.85).

The advantages of this technique are that it provides reliable results that appear to promote heart disease and can reveal a primary-secondary factor relationship that leads to an increased likelihood of heart disease.

One of the disadvantages of this technique is that it filters out a lot of detail. While those items didn't promote heart disease, they had some degree of correlation.

4.4 Fishbone Diagram Derived from C4.5 Technique

When the heart disease dataset was processed by using C4.5 technique with WEKA software, the leaf node had at least 10 elements and can be converted into a fishbone diagram as shown in Fig. 3.

From Fig. 3, the Cholesterol <= 0 and Cholesterol > 0 elements are not reasonable compared to the clustering results in Table 1 at Cholesterol = 182.63 in heart disease group and Cholesterol = 217.97 in non-heart disease group. Therefore, this element was screened out in Step 7 of integration of fishbone diagram.

The advantages of the fishbone diagram derived from decision tree is that the complex hierarchical correlation with heart disease offers more details compared to that derived from association rule tree reasoning.

The disadvantage of this method is that every hierarchical correlation between an item group and a class always appears as a root node. Heart disease was affected by several main factors, not just ST_Slope test results. Furthermore, the value of the numerical range to be used in the tree was algorithmically determined from the remaining datasets at each step of the tree creation. This makes the assignment result, although it makes the class prediction result more accurate (e.g., cholesterol level), is unreasonable when used to describe the relationship between that the attributes and the class for heart disease patients.

4.5 Fishbone Diagram Derived from Integration of All Data Mining Techniques Used in This Study

The results of parts 1–4, can be integrated into a fishbone diagram in accordance with the 12 steps of the integration algorithm as shown in Fig. 4.

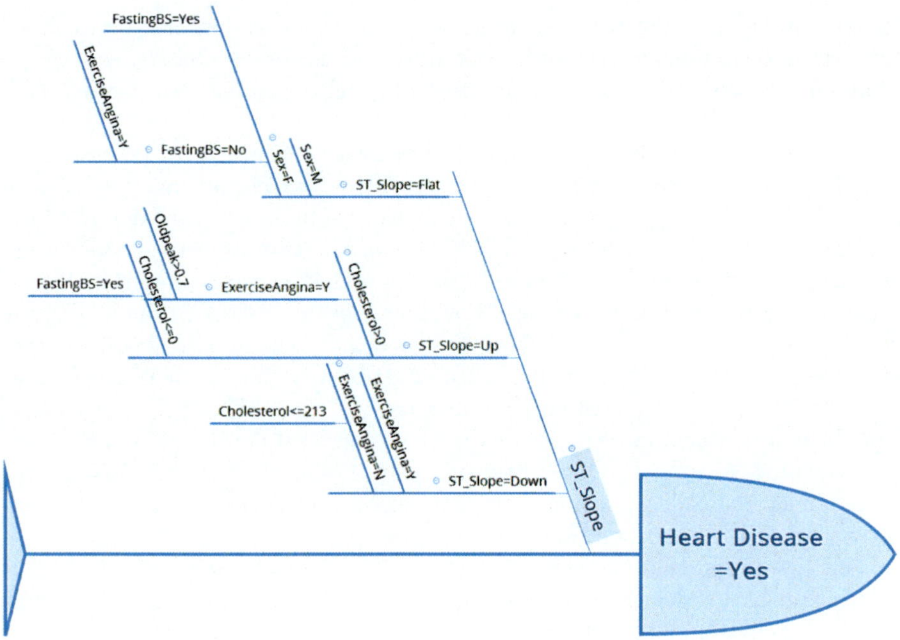

Fig. 3. Fishbone diagram derived from C4.5 technique.

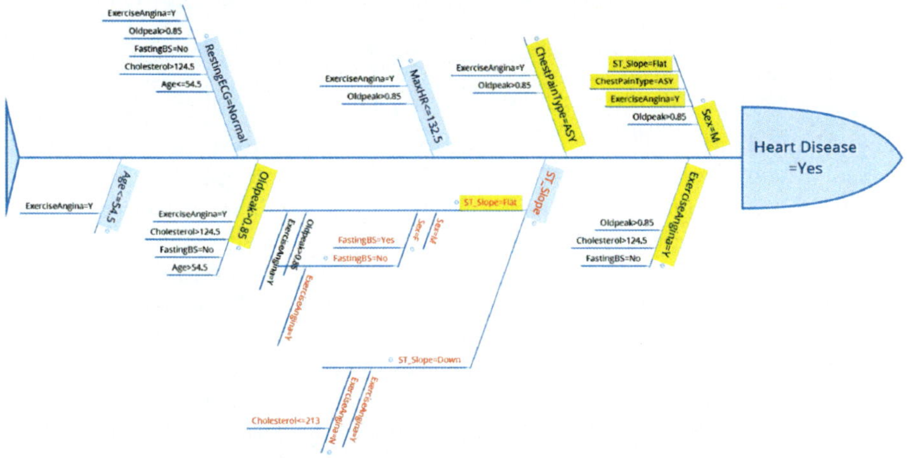

Fig. 4. Integration results of fishbone diagrams derived from data mining techniques used.

In Fig. 4, the fishbone diagram elements highlighted in yellow is the elements derived from the association rule tree. The elements in red letters are those derived from the C4.5 decision tree, where these elements showed a primary-secondary factor relationship that led to an increased likelihood of heart disease.

The remaining elements of the fishbone diagram are those obtained by the association rule technique at a minimum support and minimum confidence levels of 0.1 and 0.9, respectively and the redundant or insignificant elements were filtered out.

This integrated fishbone diagram combines the advantages of all three fishbone diagrams derived from the Apriori algorithm, association rule tree reasoning, and C4.5 techniques. The strength of the fishbone diagram derived form association rule tree reasoning is that the combination of primary-secondary factors in the fish bones with Sex = M as the main factor promotes the hierarchical correlation with heart disease. For the advantage of the fishbone diagram derived from C4.5 technique, ST_Slope showed hierarchical correlation to other items, which revealed the complexity of contributory factors of heart disease. For the highlight of the fishbone diagram derived from the Apriori algorithm, it provides a more comprehensive description of several factors of heat diseases. The advantages of the fishbone diagram derived from decision tree is that the complex hierarchical correlation with heart disease offers more details compared to that derived from association rule tree reasoning.

This integrated fishbone diagram mitigates the drawbacks of each technique by reducing excessive number of association rules with less details that is a rough grouping of environment factors. It also eliminates the drawbacks of the fishbone diagram derived from association rule tree reasoning which provides too little detail. In addition, this integrated fishbone diagram reduces the drawbacks from the C4.5 technique where the main point appears as only one root node.

5 Discussion

The fishbone diagram [9] is a tool for cause and effect analysis that is popularly applied with 4M1E topics, or Man, Machine, Material, Method, and Environment [13], for example, the analysis of the causes of accidents in factories or car accidents, etc. All sub-topics are defined for each topic, and sub-topics within each sub-topic may be defined orderly. Any topic may be defined by the user.

This research uses the fishbone diagram to analyze the causes related to cardiovascular diseases. All topics and sub-topics are defined by the knowledge discovery process with many loops of descriptive and predictive data mining techniques. Thus, the complete topics depend on the quality of the dataset. Because the dataset owner determines that each attribute in the dataset is associated with cardiovascular diseases. Therefore, we assign topics in the fishbone diagram by attributes and attribute values (or items) discovered by data mining techniques. The quality of the fishbone diagram discovered from each data mining technique can be improved by integrating fishbone diagrams, and the improvement can be understood by comparing the strengths and weaknesses of each fishbone diagram discovered from data mining techniques with the integrating fishbone diagram.

The C.4.5 Tree Classification [8] creates the tree model that attributes the order relation to the classes. The tree can be transformed into the fishbone diagram with the useful order relation, but all order relations must have the

Table 2. Comparison of fishbone diagrams generated by different algorithms.

Issues	CARs	Reasoning Framework	C4.5	Integrating
Number of topics and sub-topics	13 topics, 116 sub-topics	5 strong topics, 3 strong sub-topics	1 topic, 3 sub-topics, 13 sub of sub-topics	4 strong topics, 1 complex topic, 3 topics, 4 strong sub-topics, 14 sub-topics, and 9 sub of complex topic
Item redundancy	High redundancy and conflict on the same topic	No redundancy	No redundancy	No redundancy
Coverage of items presentation	Extensive presentation of both topics and sub-topics.	Comprehensive presentation at the topic level, but with very little sub-topics detailing.	Only one topic, but a comprehensive sub of complex topic details.	Comprehensive presentation of topics, sub-topics and sub of complex topics.
Depth of relationship levels in the fishbone diagram.	2	2	5	5
Parameter sensitivity	Depends on support and confidence	Low-sensitivity parameters	Depends on tuning parameters of C4.5	In strong topics and strong sub-topics, it has lasting results. The other part depends on adjusting the parameters of the techniques used to integrate fishbone diagrams.
Reliability of results	The higher the support and confidence, the more reliable.	A reasoning algorithm to select items that produce only one result that does not conflict is reliable.	The greedy algorithm has heuristics that are reliable in the results.	The fishbone diagram integrated from the algorithm yields reliable results.

root node. Because of this feature, its fishbone diagram only has one topic. This limitation can be overcome by the integrating fishbone diagram that retains the useful order relation from the C4.5 Tree Classification.

The association rule discovery [6] and the class association rules, or CARs [8], can be transformed from the relations between itemsets in each rule to a fishbone diagram. But these techniques generate lots of rules even with the high support and confidence values. Thus, the related items are covered, but there are lots of redundant or conflicting items. These characteristics are solved by the integrating process of our fishbone diagram.

The association rule tree [7] generates the order relationships of items related to cardiovascular disease, which can then be transformed into a fishbone diagram. But only the strongest rules with promoted order relations are discovered. This characteristic leads to the creation of the fishbone diagram with a small number of topics and sub-topics. This limitation can be solved by the integration of the fishbone diagram with topics and sub-topics discovered from other techniques that retain the strength order relation from the association rule tree.

The k-mean clustering technique is not used to generate the relationship between attribute values (or items), but rather to locate the center of each attribute value in each cluster. If each cluster has the most members corresponding to class members, cardiovascular disease or not, the center of attribute values is usable to define the reasonable attribute values (or items) of the integrating fishbone diagram, combined from many techniques where these attribute values (or items) may be conflicted values.

The comparisons between these techniques and our integrating fishbone diagram show how useful it is to decrease the limitations and to retain the strengths of these techniques. The Comparison of fishbone diagrams generated by different algorithms is shown in Table 2.

6 Conclusion

The goal of this study is to apply a variety of descriptive and predictive data mining techniques to the knowledge discovery process in order to describe the causes of cardiovascular diseases. These techniques are k-Mean Clustering, the Association Rules Discovery from Apriori Algorithm, the Association Rule Tree Reasoning, and the C4.5 tree classification. All techniques are combined by the Integrating Fishbone Diagram.

The Integrating Fishbone Diagram was discovered from 918 records in the Cardiovascular Disease Dataset. The diagram consists of 8 topics. 5 topics have the strength relation, and 2 topics have the ordered relation related to cardiovascular disease with the 16 environmental attribute values or items.

The Integrating Fishbone Diagram is created by many data mining techniques with looping on the knowledge discovery process. But this result is not complete without a bit more of the cardiovascular disease experts' review process to conclude that this is the knowledge. However, the process of integrating fishbone diagrams is an interesting example of looping the data mining processes to decrease the limitations and to retain the strengths of the fishbone diagrams discovered from these related techniques.

Acknowledgments. The authors thank the support of Department of Computer and Information Science, Faculty of Science and Engineering, Kasetsart University Chalermphrakiat Sakon Nakhon Province Campus, Thailand.

References

1. World Health Organization (WHO): Cardiovascular diseases (CVDs) (2021). https://www.who.int/news-room/fact-sheets/detail/cardiovascular-diseases-(cvds). Accessed 11 June 2021

2. Prachachat: Statistics of Thai people, an average of seven people die from heart disease every hour (2021). https://www.prachachat.net/spino/health/news-771696. Accessed 29 Sept 2021
3. Fayyad, U., Piatetsky-Shapiro, G., Smyth, P.: The KDD process for extracting useful knowledge from volumes of data. Commun. ACM **39**(11), 27–34 (1996). https://doi.org/10.1145/240455.240464
4. Han, J., Pei, J., Kamber, M.: Data mining: concepts and techniques. Elsevier (2011). https://doi.org/10.1016/B978-0-12-381479-1.00001-0
5. Fedesoriano: Heart failure prediction dataset (2021). https://kaggle.com/fedesoriano/heart-failure-prediction. Accessed 29 Sept 2021
6. Agrawal, R., Srikant, R., et al.: Fast algorithms for mining association rules. In: Proceedings of the 20th International Conference on Very Large Data Bases, VLDB, vol. 1215, pp. 487–499. Citeseer (1994)
7. Liewlom, P.: Alternative rule reasoning: association rule tree reasoning with a constraining rule ascertained using a reasoning framework in 2d interestingness area. IAENG Int. J. Comput. Sci. **48**, 3 (2021)
8. Quinlan, J.R.: C4.5: Programs for Machine Learning. Ebrary Online. Elsevier Science (2014)
9. Ishikawa, K., Loftus, J.H.: Introduction to Quality Control, vol. 98. Springer, Dordrecht (1990)
10. Holmes, G., Donkin, A., Witten, I.H.: WEKA: a machine learning workbench. In: Proceedings of ANZIIS 1994-Australian New Zealand Intelligent Information Systems Conference, pp. 357–361. IEEE (1994). https://doi.org/10.1109/ANZIIS.1994.396988
11. Liu, B., Hsu, W., Ma, Y., et al.: Integrating classification and association rule mining. In: KDD 1998: Proceedings of the Fourth International Conference on Knowledge Discovery and Data Mining, vol. 98, pp. 80–86 (1998)
12. Yun, Z.: The study of CDM-BSC-based data mining driven-fishbone applied for data processing. In: 2015 IEEE International Conference on Signal Processing, Communications and Computing (ICSPCC), pp. 1–5 (2015). https://doi.org/10.1109/ICSPCC.2015.7338909
13. Xu, Z., Dang, Y., Munro, P.: Knowledge-driven intelligent quality problem-solving system in the automotive industry. Adv. Eng. Inform. **38**, 441–457 (2018). https://doi.org/10.1016/j.aei.2018.08.013

A Self-adaptive Differential Evolution Algorithm for Solving Optimization Problems

Irfan Farda and Arit Thammano(✉)

Computational Intelligence Laboratory, Faculty of Information Technology, King Mongkut's Institute of Technology Ladkrabang, Bangkok 10520, Thailand
64607042@kmitl.ac.th, arit@it.kmitl.ac.th

Abstract. This research proposes a novel self-adaptive differential evolution algorithm for solving continuous optimization problems. This paper focuses on redesiging the self-adaptive strategy for the mutation parameters. The new mutation parameters adjust themselves to the current situation of the algorithm. When the search is stagnant, the first mutation parameter that scales the difference between the best vector and the target vector will be increased. In contrast, the second mutation parameter that scales the difference between two random target vectors will be decreased. On the other hand, when the search progresses well towards the global optimum, the algorithm will enhance the search of the surrounding space by doing the opposite of the above actions. The performance of the proposed self-adaptive differential evolution algorithm was evaluated and compared with the classic differential evolution algorithm on 7 benchmark functions. The experimental results showed that the proposed algorithm converged much faster than the classic differential evolution algorithm on all benchmark functions.

Keywords: Differential evolution · Optimization · Self-adaptive · Mutation strategy

1 Introduction

Optimization is very important in the decision-making process to choose the best solution from several alternative solutions [1]. One of the algorithms that is powerful, simple and efficient for solving optimization problems is a Differential Evolution (DE) algorithm [2, 3]. The first DE, which is called in this paper as a classic DE algorithm, introduced by Storn and Price in 1997 [4]. The DE algorithm is suitable for solving many varieties of optimization problems, such as continuous optimization problems [5, 6], large scale optimization problems [7], discrete optimization problems [8], global optimization problems [9], multiobjective optimization problems [10], combinatorial optimization problems, constrained optimization problems [11] and unconstrained optimization problems [12]. Furthermore, the DE can solve problems in many fields, including image processing citation, industrial engineering [13], electrical engineering [14], manufacturing design [15] and mechanical engineering [11].

P. Meesad et al. (Eds.): IC2IT 2022, LNNS 453, pp. 68–76, 2022.
https://doi.org/10.1007/978-3-030-99948-3_7

To increase the performance of the DE algorithm, many researchers try to modify the mutation and crossover processes because both are essential processes in the DE algorithm to generate a new individual in the population [1]. Self-adaptive control parameters become a trend. Qin and Suganthan [2] proposed a self-adaptive differential evolution for numerical optimization called SaDE. In each iteration, the control parameters were adjusted according to the model performance. Twenty-five benchmark functions were used to evaluate the algorithm performance with varying complexity. The results showed that SaDE could find the global optimal solution for several functions.

Zhang and Sanderson [16] presented a new differential evolution algorithm called JADE. The researchers introduced a new mutation strategy, DE/current-to-*p*-best, and self-adaptive control parameters. Thirteen benchmark problems were used to evaluate the algorithm performance. The simulation showed that the JADE algorithm had better convergence performance in ten benchmark problems.

Zhang and Sanderson [17] improved JADE with the optional external archive. The performance of JADE with optional external archive was compared with that of the other 6 DE variants, including the original JADE algorithm, using thirteen benchmark functions. The simulation results showed that the JADE algorithm with an optional external archive had a competitive edge for the reliability and convergence rate.

Islam et al. [18] introduced a new mutation strategy and two adaptive control parameters. They proposed the DE/current-to-gr_best/1 mutation strategy. The experimental results on twenty-five standard benchmarks and one real-life optimization problem showed that the new proposed strategy could improve the basic ability of DE algorithm to locate solution accurately in the large search space. Moreover, if integrated with the new proposed algorithm, other variants of the DE algorithm could also significantly improve.

Sun et al. [19] presented an adaptive differential evolution with a combined strategy called CSDE. They introduced two different mutation operators based on JADE algorithm. Thirty benchmark functions and four real-life problems were used to verify the proposed CSDE algorithm. The CSDE algorithm had the best performance in comparison to seven DE variants.

In this paper, we propose a new self-adaptive DE algorithm. Our algorithm uses the mutation strategy presented in Jitkongchuen and Thammano [20]. There are 2 mutation parameters in the mutation strategy: (i) if the algorithm gets stagnant, the value of the first mutation parameter will be increased while the value of the second mutation parameter will be decreased and (ii) if the algorithm keeps improving itself, the value of the first mutation parameter will be decreased while the value of second mutation parameter will be increased.

This paper is organized as follows: Sect. 2 explains the classic DE algorithm. The detail of the proposed self-adaptive DE algorithm is explained in Sect. 3. Section 4 reports the experimental results. In the end, the conclusion is discussed in Sect. 5.

2 Differential Evolution Algorithm

The Differential Evolution (DE) algorithm has four stages: initialization, mutation, crossover and selection. A brief description of the process is discussed in following:

2.1 Initialization

The first stage is the initialization. Like other evolutionary algorithms, the DE algorithm requires to specify the parameter values: the population size, the mutation parameter, the crossover parameter, and the maximum number of iterations. Then the target vectors $X_i^G = \left(x_{1i}, x_{2i}, ..., x_{ji}, ..., x_{Ni}\right)$ are randomly generated, where $i = 1, 2, ..., NP$; $j = 1, 2, ..., N$; NP is the number of target vectors in the population; N is the dimension of the target vector and the superscript G signifies the Gth generation. Note that it is necessary to adjust each element in the target vector to be within its upper and lower bounds by using the following equation:

$$x_{ji} = \min_j + rand[0, 1] \times \left(\max_j - \min_j\right) \quad (1)$$

2.2 Mutation

The mutation is the process of modifying a target vector. The result of the mutation is used as one of the vectors in the crossover process. The DE algorithm has several strategies in carrying out the mutation process. The widely used ones are DE/rand/1 and DE/current-to-best/1. The equation of DE/rand/1 is:

$$V_i^G = X_{r1}^G + F(X_{r2}^G - X_{r3}^G) \quad (2)$$

where $X_{r1}^G, X_{r2}^G, \textit{and}\ X_{r3}^G$ are three different target vectors randomly selected from. the population and F is the mutation parameter in the range of 0 to 1. The mutation parameter is used to control the amplification of the differential variation.

For DE/current-to-best/1, the mutant vector is created by using the following equation:

$$V_i^G = X_i^G + F(X_{best}^G - X_i^G) + F(X_{r1}^G - X_{r2}^G) \quad (3)$$

where X_{best}^G is the target vector with the best fitness value in the Gth generation.

2.3 Crossover

After the mutation process, the crossover is carried out to produce offspring $U_i^G = \left(u_{1i}, u_{2i}, ..., u_{ji}, ..., u_{Ni}\right)$, called trial vectors. The crossover combines the target vector with the mutation vector based on the following equation:

$$u_{ji} = \begin{cases} v_{ji};\ \textit{if rand} \le CR \textit{ or } j = rd \\ x_{ji};\ \textit{otherwise} \end{cases} \quad (4)$$

where rand is a random number in the range [0, 1]; CR is the crossover parameter. CR is a recombination probability. The trial vector element is selected from the mutant vector if rand is less than or equal to CR. Otherwise, it is selected from the target vector. rd, a random integer in the range [1, N], is used to ensure that at least one element of the trial vector is different from the target vector.

2.4 Selection

The last stage is the selection. This stage determines the new population for the next generation. The selection process considers the fitness values of the trial and target vectors. The better fitness between the trail vector and the target vector is selected to fill the new population.

$$X_i^{G+1} = \begin{cases} U_i^G; f(U_i^G) \le f(X_i^G) \\ X_i^G; \ otherwise \end{cases} \tag{5}$$

The algorithm runs 2.2 to 2.4 repeatedly until any of the stopping conditions is satisfied.

3 Proposed Method

Our proposed method is based on a self-adaptive differential evolution algorithm by Jitkongchuen and Thammano [20] that modified the DE/current-to-best/1 mutation strategy to contain two mutation parameters, λ and F, as shown in (6).

$$V_i^G = X_i^G + \lambda(X_{best}^G - X_i^G) + F(X_{r1}^G - X_{r2}^G) \tag{6}$$

where λ is the first mutation parameter for scaling the difference between the best vector and the target vector; X_{best}^G is the best fitness vector in the Gth generation. F is used for scaling the difference between two random target vectors. The two mutation parameters are adjusted in each generation according to:

$$\lambda^{G+1} = \lambda^G - \left(\lambda^G \times C_1\right) \tag{7}$$

$$F^{G+1} = F^G + \left(F^G \times C_2\right) \tag{8}$$

where C_1 and C_2 are a uniform random number in the range [0, 0.01] and [0, 1] respectively. In each subsequent iteration, the λ value is decreased to decrease the effect of the best vector while the value of F is increased to increase the impact of the random vector. The algorithm is able to avoid being trapped in the local minimum with this combination.

This paper proposes new equations to adjust λ and F. The new Eqs. (9) and (10), adjust λ and F according to the current situation of the algorithm.

$$\lambda^{G+1} = \begin{cases} \lambda^G + \left(\lambda^G \times C_1\right); \ if\ stagnation\ happens \\ \lambda^G - \left(\lambda^G \times C_1\right); \ otherwise \end{cases} \tag{9}$$

$$F^{G+1} = \begin{cases} F^G - \left(F^G \times C_2\right); \ if\ stagnation\ happens \\ F^G + \left(F^G \times C_2\right); \ otherwise \end{cases} \tag{10}$$

where C_1 and C_2 are tuning parameters that control the step size. If the algorithm is in stagnation state, the value of λ will be increased and the value of F will be decreased. Otherwise, to enhance the global search effect, the value of λ will be decreased and the value of F will be increased. Note that the values of λ and F must be in the range of 0 to 1. If the values exceed 1, the algorithm will set them back to 1. If the values fall below 0, the algorithm will set them back to 0.

4 Experimental Results

Table 1 shows 7 benchmark functions based on Yao et al. [21]. These benchmark functions are used to evaluate the performance of the proposed self-adaptive DE algorithm. The functions are divided into 2 categories: the functions f_1–f_5 are continuous unimodal functions, and the functions f_6 and f_7 are multimodal functions.

Table 1. The benchmark functions.

No.	Function name	Equation
f_1	Sphere model	$f_1(x) = \sum_{i=1}^{N} x_i^2, N = 30, S = [-100, 100], f_{\min} = 0$
f_2	Schwefel's problem 2.22	$f_2(x) = \sum_{i=1}^{N} \lvert x\rvert + \prod_{i=1}^{N} \lvert x\rvert, N = 30, S = [-10, 10], f_{\min} = 0$
f_3	Schwefel's problem 1.2	$f_3(x) = \sum_{i=1}^{N} \left(\sum_{j=1}^{i} x_j\right)^2, N = 30, S = [-100, 100], f_{\min} = 0$
f_4	Schwefel's problem 2.21	$f_4(x) = \max_i\{\lvert x_i\rvert, 1 \le i \le N\}, N = 30, S = [-100, 100], f_{\min} = 0$
f_5	Generalized Rosenbrock's	$f_5(x) = \sum_{i=1}^{N-1} \left[100\left(x_{i+1} - x_1^2\right)^2 + (x_i - 1)^2\right], N = 30, S = [-30, 30], f_{\min} = 0$
f_6	Shekel's foxholes	$f_6(x) = \left[\frac{1}{500} + \sum_{j=1}^{25} \frac{1}{j + \sum_{i=1}^{N} (x_i - a_{ij})}\right]^{-1}, N = 2, S = [-65.539, 65.536], f_{\min} = 0.998004$
f_7	Kowalik's	$f_7(x) = \sum_{i=1}^{11} \left[a_i - \frac{x_i(b_i^2 - b_i x_2)}{b_i^2 + b_1 x_3 + x_4}\right]^2, N = 4, S = [-5, 5], f_{\min} = 0.0003075$

This paper compares the performance of the proposed self-adaptive DE algorithm with the classic DE algorithm. The comparison is performed in terms of the convergence speed and the ability to find the global optimum. To compare the results of the proposed algorithm with those of the classic DE, the experiments are conducted in the same way for both algorithms: (i) for each test function, each algorithm is run 10 times with different initial populations each time and (ii) the control parameters are set to the same values as shown in Table 2.

Table 2. The control parameters.

Parameters	Values
Population size (NP)	100
Mutation parameters	
$\lambda(0)$	0.5
$F(0)$	0.5
Crossover parameter (CR)	0.9
C_1	0.01
C_2	0.02

Figure 1 shows the convergence performance of the proposed self-adaptive DE and the classic DE algorithms on all 7 benchmark functions. We can see that the proposed self-adaptive DE algorithm has better convergence performance than the classic DE. The proposed self-adaptive DE algorithm rapidly rolls down towards the global optimum on all functions while the classic DE algorithm requires more iterations to converge to the global optimum. Furthermore, the proposed self-adaptive DE algorithm can achieve global or near-global optimum before the maximum number of iterations is reached. As can be seen in Fig. 1, the classic DE algorithm got stagnant very frequently throughout the search while the proposed self-adaptive DE algorithm did not. This improved performance is mainly due to the effectiveness of the self-adaptive strategy. More precisely, in the early stage of the search, the self-adaptive DE algorithm accelerates the search by increasing F to increase the exploration capability and decreasing λ to decrease the exploitation capability. On the other hand, when the algorithm is in stagnation state, the algorithm tries to escape from it by decreasing F to decrease the exploration capability and increasing λ to increase the exploitation capability.

Table 3 compares the ability to find the global optimum of the proposed DE algorithm with the classic DE algorithm. There are two numbers in each cell of the table. The first is the mean result over 10 repetition runs obtained by each algorithm. The second number, which is in parentheses, represents the standard deviation over 10 repetition runs. The experimental results show that both algorithms perform equally well in terms of the ability to find the global optimum. They can achieve global optimum on the functions f_5 and f_6, and achieve near-optimum on f_1, f_2, f_3, f_4, f_7.

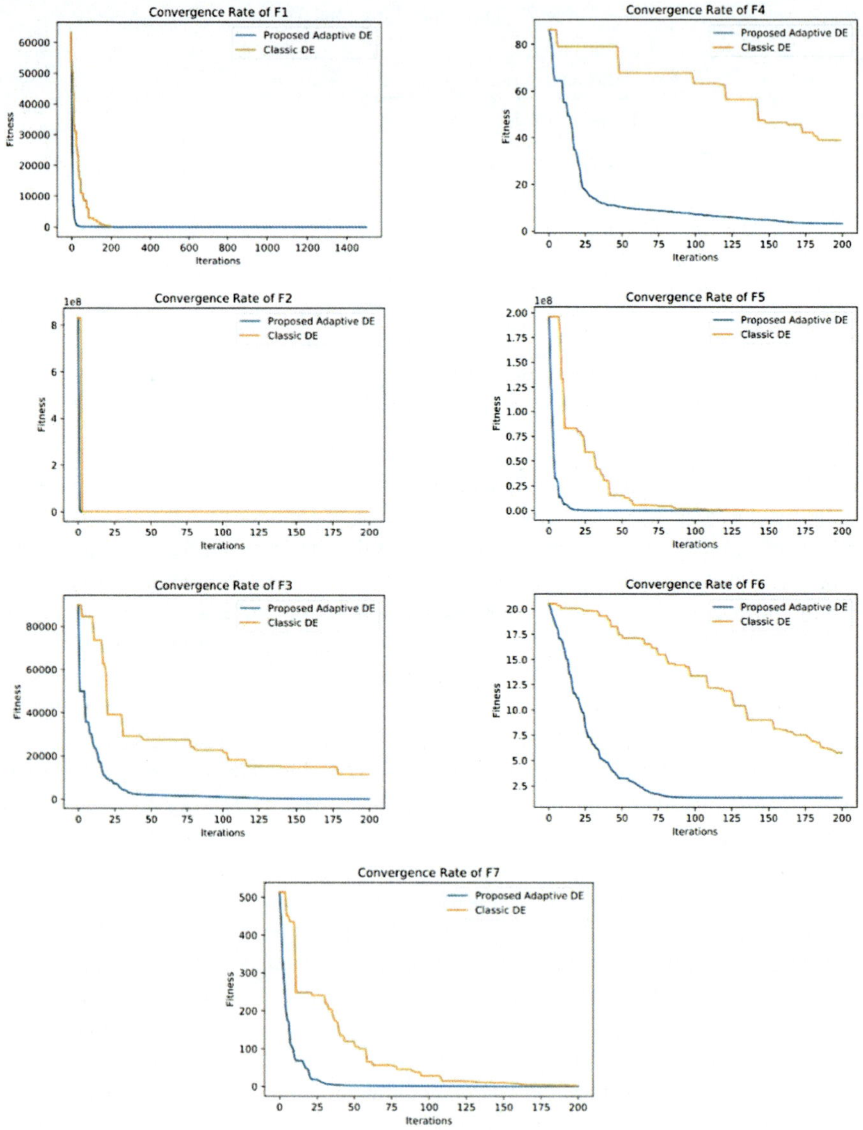

Fig. 1. Comparative convergence rate of the classic and the proposed self-adaptive Des.

Table 3. Comparative results of the classic DE and the proposed self-adaptive DE.

Functions	Iterations	Classic DE	Proposed Self-adaptive DE
f_1	1500	2.41E-14 (2.61E-08)	2.41E-58 (2.80E-30)
f_2	2000	1.34E-10 (6.31E-12)	4.63E-40 (2.60E-41)
f_3	5000	4.78E-12 (4.38E-07)	1.18E-38 (3.47E-20)
f_4	5000	5.71E-10 (3.43E-10)	3.31E-05 (2.40E-05)
f_5	20000	0 (0)	0 (0)
f_6	100	0.998004 (0)	0.998004 (0)
f_7	4000	0.0051772 (0.07)	0.0051772 (0.0691)

5 Conclusion

In this paper, we presented a self-adaptive differential evolution (DE) algorithm for solving continuous optimization problems. Based on DE/current-to-best/1 strategy, we modified the mutation parameters to have an ability to self-adapt to the current situation of the algorithm. The proposed algorithm has been evaluated and compared with the classic differential evolution algorithm on 7 benchmark functions. The experimental results showed that both algorithms performed equally well in terms of the ability to find the global optimum. However, the proposed algorithm has better convergence performance than the classic differential evolution algorithm on all benchmark functions. In the future work, we would like to introduce a variable step size technique to the mutation parameter adjustment equations.

Acknowledgement. This work was supported by King Mongkut's Institute of Technology Ladkrabang.

References

1. Bilal, Pant, M., Zaheer, H., Garcia-Hernandez, L., Abraham, A.: Differential evolution a review of more than two decades of research. Eng. Appl. Artif. Intell. **90**, 103479 (2020)
2. Qin, A.K., Suganthan, P.N.: Self-adaptive differential evolution algorithm for numerical optimization. In: 2005 IEEE Congress on Evolutionary Computation, IEEE CEC 2005. Proceedings, vol. 2, pp. 1785–1791. IEEE, Edinburgh (2005)
3. Zeng, Z., Zhang, M., Chen, T., Hong, Z.: A new selection operator for differential evolution algorithm. Knowl.-Based Syst. **226**, 107150(2021)
4. Storn, R., Price, K.: Differential evolution – a simple and efficient heuristic for global optimization over continuous spaces. J. Glob. Optim. **11**(4), 341–359 (1997). https://doi.org/10.1023/A:1008202821328
5. Yi, W., Zhou, Y., Gao, L., Li, X., Mou, J.: An improved adaptive differential evolution algorithm for continuous optimization. Expert Syst. Appl. **44**, 1–12 (2016)
6. Zhou, Y.Z., Yi, W.C., Gao, L., Li, X.Y.: Adaptive differential evolution with sorting crossover rate for continuous optimization problems. IEEE Trans. Cybern. **47**, 2742–2753 (2017)

7. Deng, W., et al.: Quantum differential evolution with cooperative coevolution framework and hybrid mutation strategy for large scale optimization. Knowl.-Based Syst. **224**, 107080 (2021)
8. Cuevas, E., Zaldivar, D., Pérez-Cisneros, M., Ramírez-Ortegón, M.: Circle detection using discrete differential evolution optimization. Pattern Anal. Appl. **14**, 93–107 (2011)
9. Yu, X., Cai, M., Cao, J.: A novel mutation differential evolution for global optimization. J. Intell. Fuzzy Syst. **28**, 1047–1060 (2015)
10. Wang, J., Zhang, W., Zhang, J.: Cooperative differential evolution with multiple populations for multiobjective optimization. IEEE Trans. Cybern. **46**, 2848–2861 (2016)
11. de Melo, V.V., Carosio, G.L.C.: Investigating multi-view differential evolution for solving constrained engineering design problems. Expert Syst. Appl. **40**, 3370–3377 (2013)
12. Zhao, Z., Yang, J., Hu, Z., Che, H.: A differential evolution algorithm with self-adaptive strategy and control parameters based on symmetric Latin hypercube design for unconstrained optimization problems. Eur. J. Oper. Res. **250**, 30–45 (2016)
13. Kadhar, K.M.A., Baskar, S., Amali, S.M.J.: Diversity controlled self adaptive differential evolution based design of non-fragile multivariable PI controller. Eng. Appl. Artif. Intell. **46**, 209–222 (2015)
14. Coelho, L.D.S., Mariani, V.C., Ferreira Da Luz, M.V., Leite, J.V.: Novel gamma differential evolution approach for multiobjective transformer design optimization. IEEE Trans. Magn. **49**, 2121–2124 (2013)
15. Kao, Y., Chen, C.C.: A differential evolution fuzzy clustering approach to machine cell formation. Int. J. Adv. Manuf. Technol. **65**, 1247–1259 (2013). https://doi.org/10.1007/s00170-012-4254-5
16. Zhang, J., Sanderson, A.C.: JADE: self-adaptive differential evolution with fast and reliable convergence performance. In: 2007 IEEE Congress on Evolutionary Computation, CEC 2007, pp. 2251–2258. IEEE, Singapore (2007)
17. Zhang, J., Sanderson, A.C.: JADE: adaptive differential evolution with optional external archive. IEEE Trans. Evol. Comput. **13**, 945–958 (2009)
18. Islam, S.M., Das, S., Ghosh, S., Roy, S., Suganthan, P.N.: An adaptive differential evolution algorithm with novel mutation and crossover strategies for global numerical optimization. IEEE Trans. Syst. Man Cybern. B Cybern. **42**, 482–500 (2012)
19. Sun, G., Yang, B., Yang, Z., Xu, G.: An adaptive differential evolution with combined strategy for global numerical optimization. Soft. Comput. **19**, 1–20 (2019). https://doi.org/10.1007/s00500-019-03934-3
20. Jitkongchuen, D., Thammano, A.: A self-adaptive differential evolution algorithm for continuous optimization problems. Artif. Life Rob. **19**(2), 201–208 (2014). https://doi.org/10.1007/s10015-014-0155-z
21. Yao, X., Liu, Y., Lin, G.: Evolutionary programming made faster. IEEE Trans. Evol. Comput. **3**, 82–102 (1999). https://doi.org/10.1007/s10015-014-0155-z

Chertify: Wood Identification-Based Mobile Cross-platform by Deep Learning Technique

Teerasak Wongpoo, Wannamongkol Sriwan, Taravichet Titijaroonroj, and Pornsuree Jamsri(✉)

Faculty of Information Technology, King Mongkut's Institute of Technology Ladkrabang, Bangkok, Thailand
{61070094,61070201,taravichet,pornsuree}@it.kmitl.ac.th

Abstract. Thailand's economic trees are counted as one of its most valuable domestic assets and well known internationally as a high quality natural wood resource. However, there is a need for basic wood identification whether or not for a required certificate by individuals, entrepreneurs, and organizations. Currently, the wood identification process is manually accomplished only by an expert at the Forest Research and Development Office, the Royal Thai Forest Department. This is a time consuming complex process for two reasons–required experience and limited experts. Given the complexity of wood identification, a new approach is offered, namely, to identify different types of wood with an image from a smartphone. The researcher initially proposes a mobile application, "Chertify", that has five features (login, wood check, wood check history, manual, and wood knowledge). This app can serve both iOS and Android platforms and targets the general user. Chertify aims to simplify identification of an economic wood type by combining deep learning technology with an actual wood image on a Smartphone. The selected deep learning algorithm will be applied to 258 trained group images of seven wood types based on highest accuracy and lowest standard deviation. Chertify relies on a handcrafted method (HOG and SVM) and a learning-based method (Alexnet) with accuracy of 69.4% and 84.73% and SD at 5.37% and 3.07% values, respectively.

Keywords: Wood identification · Deep learning · Machine learning · Mobile application

1 Introduction

Significantly, Thailand's existing forest area is approximately 163,765 sq. meters or 31.64% nationwide since 2020 [1]. In 2020, the Forest Certification Section Bureau of Forestry Economy, Royal Forest Department also indicated that the value of certified wood, wood products and charcoal for trade or export outside the country, fiscal year 2020 was 92.84 million USD (100%). The majority of

P. Meesad et al. (Eds.): IC2IT 2022, LNNS 453, pp. 77–87, 2022.
https://doi.org/10.1007/978-3-030-99948-3_8

wood (logs) and wood products (lumber, sawn timber, wood in chips or particles) exported of 5 million cubic meters is worth 92.61 million USD (75%) [1]. Wood products export came mainly from the economic tree as it's high in quality and value. In 2019, an annual report of the Forest Research and Development Office (FRDO) stated they tested 657 cases for wood identification including certified wood type and its quality [2]. Nevertheless, there are 96 wood experts consisting of 64 active staff, and 32 retired staff who were relocated to other organizations [3]. Each piece of wood investigation takes about 7–9 days. Comparing the number of cases and active staff is equal to 1:10. This implies that one staff person has to inspect 10 pieces per year.

Basically, the conventional wood identification method uses a hand lens or a microscope to look directly at a piece of wood. The expert can specify wood type with those tools combined with knowledge and experience with the wood's anatomy. Some woods, however, are difficult to inspect relying solely the on human eye [4]. Two major goals of this research are to help a general user to be able to identify wood correctly within a reasonable amount of time without concern about a wood certificate and to reduce the workload of FRDO staff at the beginning identification stage. A mobile application with cutting-edge technology is introduced by merging with a deep learning algorithm. Deep learning technology enables a mobile app to identify the wood type. The Royal Forest Department has contributed 7 types of dataset out of 15 types for this research. At this initial stage, the prototype started with 7 types of 258 wood's images totally. Although, it required the trained dataset and high algorithm to utilize accurate results in the application, there are various available algorithms, namely a handcrafted method or learning-based methods [5–12]. This, then, becomes another alternative for target groups both a general user and an expert at FRDO [13]. Accordingly, it can serve mass users with an affordable price after an acceptable development period.

To serve beginner users, Chertify App is developed with five features: (i) Login (ii) Wood identification (iii) Wood identification history (iv) Manual (v) Wood knowledge. It is available for cross platforms-iOS and Android users. Thus, users can employ their own available smartphone to gather the wood's image to identify wood type by using the deep learning algorithms. This paper includes 6 parts: (i) introduction; (ii) related work; (iii) system design; (iv) wood identification methodology; (v) experiments and discussion; and vi) conclusion.

2 Related Work

2.1 Economic Wood Situation in Thailand

The land area of Thailand is 510,890 sq.kms with a forest area of 198,730 sq.kms (38.89 % of total land) as reported by the World Bank in 2020 [14]. Economically, agriculture, forestry and finishing valued added is 9% of GDP and exports of goods and services is 51% of GDP as stated by World Development Indicators database in 2020 [15]. Although the forest area was reduced 1,998 sq.kms from the previous decade, wood products are still the majority of exported products of Thailand's economy. In particular, export of certified sawn wood was worth 29 thousand million baht of 2.3 million cubic meters in 2020 as reported by Thailand's Customs Department [1]. A significant part of exported wood products data is to identify the type of economic wood officially for entrepreneurs, wood industry, governmental organizations, and individuals. However, there is also another portion of people that need to know a wood's type without requiring an official certificate for personal reasons such as to check a type of existing old furniture, ensure the type before purchasing or selling it to or from a company, and to check the wood heritage.

2.2 Conventional Wood Identification Process

Usually, the wood identification process is done by FRDO. This wood is represented as a hardwood category. Recently, they checked 657 cases for wood identification including certifying wood type and wood characteristics, and its quality [2]. With a hand lens 10× to 15× through a cross section and wood surface, the general characteristics of Thai timbers besides focus on outside appearance (color, heartwood, sapwood, wood splinter) but also on 3 functions (i) the conducting tissue (rays or pores), (ii) the strengthening tissue (wood fibers), and (iii) the storage tissue (wood parenchyma, ray parenchyma cell) [4,16]. In 2020, the total of 96 wood experts were 64 active staff, and 32 retired staff relocated to other organizations [3]. The experts are not only in charge of wood identification but also responsible for other major work at FRDO in the office and onsite in the forest. The staff realizes the process requires time even though the steps are simple in their conventional identification.

Staff does inspection manually with two things 1) a hand lens of 10× to 15× magnification and 2) a cutter knife. Then, they do these four steps (i) select the right position to cut, (ii) cut a piece of wood to get the cross section and wood surface, (iii) visually explore the wood's characteristics through the hand lens, (iv) identify the wood type. This is done by an expert who knows the right spot to cut that does not damage the sample too much. With experience and knowledge about economic wood, staff is able to recognize the wood type easily.

2.3 Wood Identification Technology and Tools

In addition, there is the advanced approach and applications that are provided to the user and staff to identify wood accurately. The advanced method needs to be performed in a wood DNA testing laboratory with advanced equipment and multiple steps to identify the actual wood type. Besides requiring expertise and specific skills, there is also a high cost that comes with the identification process. However, this research will bring an affordable wood identification procedure in terms of time, skill, equipment, and price for conventional wood identification. For example, the wood DNA testing laboratory gives reliable results with a high accuracy rate, is fast yet for an expensive price. The DNA testing of timber is the choice for enterprises and important organizations but unsuitable for a general user. Double Helix Tracking Technologies Pte Ltd. offers wood DNA testing commercially so that their clients can trace the products back to origin [17].

For technology, there is previous research applying machine learning technology (HOG for feature extraction and SVM for classification) with a 70.5% positive testing result for wood images to identify wood's types among 4,000 wood species in Indonesia [5]. Another similar research uses GLCM algorithm for feature extraction and SVM for classification of 24 wood species for wood recognition with 91.7% accuracy. Both have the same goal, namely, to simplify inspection steps and utilize technology for wood identification effectively [6].

Other recent techniques, namely, feature extraction (HOG), classification (SVM) and reduction of wood image size but not quality (Gaussian Pyramid), laplacian ledge detection (distinguish pattern), are applied on a mobile app-Android platform. The Indonesian's research team illustrated that these algorithms improved accuracy (lowest increment accuracy 4%) of wood identification on 5 wood species dataset of Indonesia [7]. The AIKO-KLHK app is another existing wood identification app available for Android Smarthphone. This app is conducted in the Indonesian language and based on Indonesian Wood with four main functions–Wood Identification, History, Library, and Guideline [13].

3 System Design

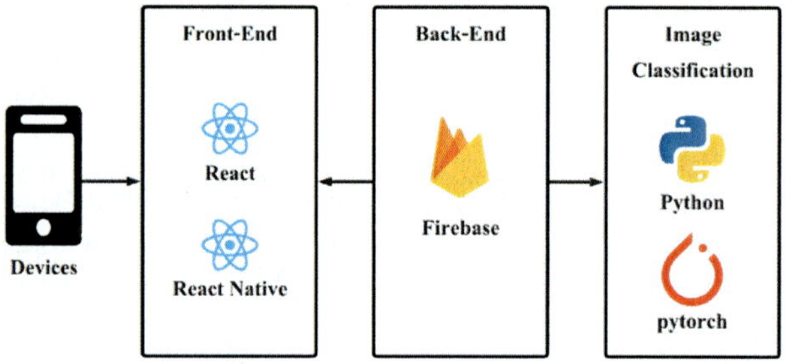

Fig. 1. System overview.

In this section, the main objective is to briefly explain the details of the Chertify application. This section is divided into three parts including (i) Front-end, (ii) Back-end, and (iii) Image Recognition as shown in Fig. 1. Firstly, our graphic user interface (GUI) is designed and developed by using Figma and React Native, respectively. These are widely used and can make our application to be a mobile cross-platform application.

Secondly, all data are stored in the Firebase store which is a part of the Firebase cloud platform. This platform is designed to support the developer for building and deploying a mobile application. As depicted in Fig. 2, the ER-diagram is used to present our schema. There are five main entities including

User, Wood, History, Content, and Announcement. Firstly, the user entity is designed to support the login feature. Secondly, the announcement and content entities are created to collect all data of the information and the user manual functions, respectively. Lastly, the wood and history entities are used together when users need to predict the given wood image. The wood information from the wood entity is displayed to users before saving its result to the history entity when the wood identification process is completed.

Lastly, to construct the image classification service, the PyTorch library, which is an open source machine learning framework, is employed. This is used to identify the wood image and return the wood information to the user.

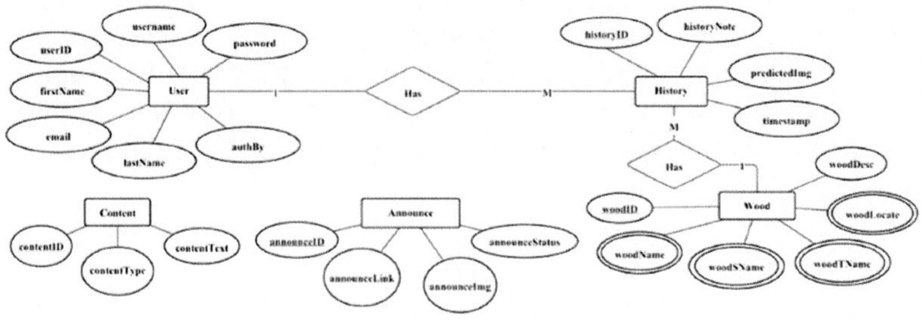

Fig. 2. An ER-diagram of the chertify application.

4 Wood Identification Methodology

In this section, we aim to explain the details of the wood identification methodology in the Chertify application. In the image recognition domain, there is much research discussing wood identification from a given image. These can be divided into two groups: (i) handcraft-based and (ii) learning-based groups as follows:

In the handcraft-based group, the image recognition task consists of two main parts of feature extraction and classification. Firstly, there are many feature extractions for wood identification such as Histogram of Oriented Gradients (HOG), Gray-Level Co-Occurrence Matrix (GLCM), and Local Binary Pattern (LBP). These are widely used for extracting the shape and texture features from the whole image. These features are used to be representative for each image. Secondly, the classification step is an important way to indicate the wood type by considering the highest probability value of class in each image. There are many existing classifications including K-Nearest Neighbor (KNN), Support-Vector Machine (SVM), and Artificial Neural Network (ANN). The most challenging part of the handcraft-based group is how to ensure that the right methods in feature extraction and classification are suitable for a specific wood identification. Generally, the decision for selecting the appropriate feature extraction and classification is based on the research experience and experiments. In the training phase, the need is to adjust the feature extraction and classification separately because the handcraft method is a multiple-stage procedure.

In the learning-based group, many methods are single-stage as it is a combination of both feature extraction and classification parts together. The methods can extract an important feature to recognize and determine the correct group for each sample. This method can learn and adjust automatically from the training data with its label. It is easy to use and does not require much experience. Recently, there are many deep learning methods available for image classification including Alexnet [8], Resnet [9], Vggnet [10], Squeezenet [11], and Inception [12]. These have been constructed from the convolution neural network (CNN) concept. This consists of three main layers–(i) convolution layer, (ii) polling layer, and (iii) fully connected layer as shown in Fig. 3. The first layer is employed to extract the rich information from the whole image. While the pooling layer is designed to reduce the feature size to make it robust to scaling, shift, rotational in-variances. This, nevertheless, can still decrease time consumption because the feature map size is reduced. Lastly, the fully connected layer is used to classify the given image to the corrected group. However, the method in the learning-based group requires more training data. Its performance depends on the variety of the training data.

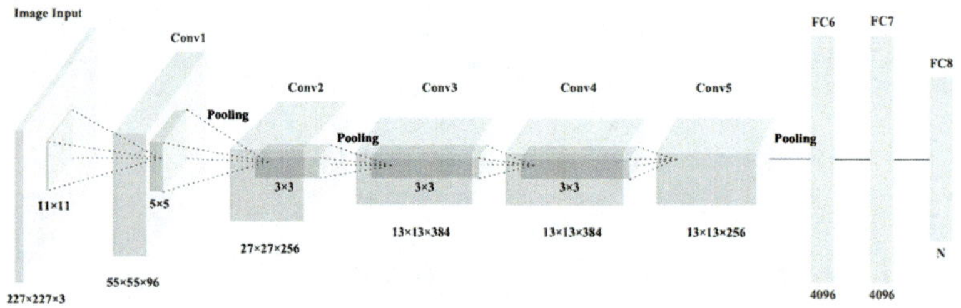

Fig. 3. An overall of the Alexnet architecture.

To achieve the high performance of the wood identification, we have experimented to compare each method in both groups by using our dataset as shown in Sect. 5.2.

5 Experiments and Discussion

5.1 Prototype of Chertify Application

Our application consists of five main features including (i) login, (ii) identify the wood image, (iii) read the wood information, (iv) study the user manual, and (v) see logging. The overviews are shown in Fig. 4.

First, the login view as displayed in Fig 4(a) aims to identify the user before entering our application. There are four ways to sign up (i) Facebook account, Google email, Line account, and Chertify account. This feature can help the user to reuse an existing account instead of creating the new one for our application. After that, the home page will appear if the login process is a success as shown in Fig 4(e).

Fig. 4. Chertify application prototype.

Secondly, the view of the wood identification is shown in Fig 4(b). Users can take a photo from the Smartphone before feeding it to the wood identification service. The wood information, which has the highest confidence score, will appear to the user.

As shown in Fig. 4(c), the wood information view consists of two main views including the table of wood and its detail. This view is proposed to give the wood information to the user by selecting the wood card or type the wood name into the text field.

Lastly, the views for studying the user manual and seeing logging are illustrated in Fig. 4(d). The designs are intended to support the users when they need to review the wood image and its detail again for seeing the logging. The background to prepare the wood before capturing its image and the user's guideline are provided in the user manual view.

5.2 Training and Testing for Wood Identification

To compare the performance of each existing method, the Thailand economic wood image (Thai-EWI) dataset, which is collected from FRDO at the Royal Forest Department in Thailand, is used. The Thai-EWI dataset contains 258 images with resolution 1200×1200 pixels. There are seven classes including 60 Iron wood images, 12 Teng images, 24 Phayom images, 18 Phayuung images, 18 Teak images, 24 Thingan images, and 102 Praduu images as shown in Fig. 5.

The k-fold cross validation is used to avoid the overfitting problem in machine learning algorithms. In this experiment, k is set to 5, equally, and the accuracy rate is used to indicate the performance in each method. The baseline method

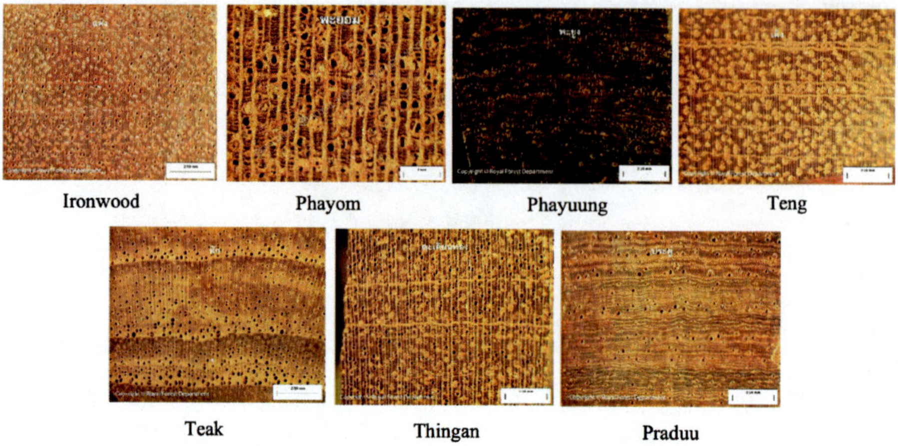

Fig. 5. Cross section of woods.

can be divided into two groups: handcraft-based and learning-based methods. In the first group, there are six methods including (i) HOG-SVM, (ii) HOG-KNN, (iii) LBP-SVM, (iv) LBP-KNN, (v) GLCM-SVM, and (vi) GLCM-KNN. In the second group, there are five methods including Alexnet [8], Resnet [9], Vggnet [10], Squeezenet [11], and Inception [12]. The configuration of the learning-based methods including learning rate, momentum, epoch, learning method, and loss function is summarized in Table 1.

Table 1. Parameters used by Alexnet, Resnet, Vggnet, Squeezenet, and Inception.

Method	Learning rate	Momentum	Epoch	Learning method	Loss function
Resnet	0.001	0.9	25	SGD*	Cross entropy loss
Alexnet	0.001	0.9	25	SGD*	Cross entropy loss
Vggnet	0.001	0.9	25	SGD*	Cross entropy loss
Squeezenet	0.001	0.9	25	SGD*	Cross entropy loss
Inception	0.001	0.9	25	SGD*	Cross entropy loss

*Stochastic Gradient Descent (SGD) is a well-known learning method.

The experimental results in Table 2 show that the overall performance in the handcraft-based method is 15.43 lower than the learning-based method when compared with HOG-SVM and Alexnet, which are the best methods in each group. The average of the handcraft-based and learning-based groups are 46.83% and 83.19%, respectively. This means that the performance of the learning-based group is higher than the handcraft-based group. For the algorithm, the accuracy rate is higher and the standard deviation remains lower. This means that the learning-based group is robust with a high performance. Therefore, the wood identification in our application is implemented by using the technique from the

Table 2. Performance of machine learning and deep learning methods.

Methods	K1	K2	K3	K4	K5	AVG ± S.D.
HOG-SVM	67.00%	67.00%	63.00%	75.00%	75.00%	69.40% ± 5.37%
HOG-KNN	63.00%	54.00%	48.00%	69.00%	65.00%	59.80% ± 8.58%
LBP-SVM	38.00%	38.00%	38.00%	39.00%	43.00%	39.20% ± 2.17%
LBP-KNN	38.00%	23.00%	23.00%	25.00%	24.00%	26.60% ± 6.43%
GLCM-SVM	48.00%	50.00%	46.00%	57.00%	61.00%	52.00% ± 6.35%
GLCM-KNN	29.00%	40.00%	33.00%	31.00%	35.00%	34.00% ± 4.22%
Resnet	**88.76%**	81.40%	81.40%	**81.78%**	83.33%	83.33% ± 3.14%
Alexnet	88.37%	81.40%	**86.82%**	**81.78%**	**85.27%**	**84.73**% ± **3.07**%
Vggnets	84.50%	**82.56%**	85.66%	81.01%	84.50%	83.64% ± 1.85%
Squeezenet	85.27%	**82.56%**	84.88%	81.01%	83.33%	83.41% ± 1.74%
Inception	84.11%	79.85%	80.23%	80.85%	80.23%	80.85% ± 1.83%

learning-based group. When we were considering the learning-based group, we found that the performance of the Alexnet is the highest. It is appropriate for the wood identification when it is tested by the Thai-EWI dataset. However, the accuracy value of the Vggnet and Squeezenet methods differ only 1% higher than the Alexnet method. This is considered as near to each other. The standard deviation values of Vggnet and Squeezenet are also only 1% lower than the Alexnet. We can conclude that the performance of the Alexnet, Vggnet and Squeezenet are near equal. For the Chertify application, we use the Alexnet for the wood image classification because its accuracy rate is currently the highest value.

6 Conclusion and Future Research

At this stage, the research team have successfully developed the prototype Chertify App with five features to serve mainly individuals. The app is based on DL–Alexnet with accuracy of 84.73% and SD at 3.07% values. The benefits of Chertify App towards our target group meets the goals as planned. Significantly, it directly serves identify wood type for general users and reduces the workload of FRDO staff.

For future research, the team will extend the app performance with utilization of the DL algorithm and useful features based on a total of 15 types of wood. When receiving additional datasets from FRDO, the development of Chertify App will continue to fulfill our goals. Then, the team can launch this mobile application to the public.

Acknowledgement. Wood image dataset and beneficial information are contributed by FRDO, the Royal Forest Department. Researcher would like to give special thanks to FRDO staff.

References

1. Royal Forest Department (2020) Forestry Statistics Data 2020. http://forestinfo.forest.go.th/55/Content.aspx?id=1040. Accessed 10 Aug 2021
2. Royal Forest Department (2019) Annual Report of 2019. http://forprod.forest.go.th/forprod/forprod2017/Knowledgebase/file/e-book/μ/μ256D2/2562-2.pdf. Accessed 10 Aug 2021
3. Royal Forest Department (2021). A list of expert in wood inspection. http://forprod.forest.go.th/forprod/forprod2017/PDF/æμ-2563-Watermark%5B31-07-63%5D.pdf. Accessed 10 Aug 2021
4. The Forest Research and Development Office: Hardwoods Identification. Kunathai Co., Ltd (2011)
5. Sugiarto, B., et al.: Wood identification based on histogram of oriented gradient (HOG) feature and support vector machine (SVM) classifier. In: 2017 2nd International Conferences on Information Technology, Information Systems and Electrical Engineering (ICITISEE). pp. 337–341. IEEE, Yogyakarta (2017). https://doi.org/10.1109/ICITISEE.2017.8285523
6. Bi-hui Wang, Wang, H., Qi, H.: Wood recognition based on grey-level co-occurrence matrix. In: 2010 International Conference on Computer Application and System Modeling (ICCASM 2010), pp. V1-269–V1-272. IEEE, Taiyuan, China (2010). https://doi.org/10.1109/ICCASM.2010.5619388
7. Sugiarto, B., Arifin, M.R., Laluma, R.H., Prakasa, E., Gunawansyah, Azwar, A.G.: An improved wood identification accuracy using gaussian pyramid and Laplacian edge detection based on android smartphone. In: 2020 14th International Conference on Telecommunication Systems, Services, and Applications (TSSA). pp. 1–5. IEEE, Bandung, Indonesia (2020). https://doi.org/10.1109/TSSA51342.2020.9310813
8. Krizhevsky, A., Sutskever, I., Hinton, G.E.: ImageNet classification with deep convolutional neural networks. Commun. ACM. **60**, 84–90 (2017). https://doi.org/10.1145/3065386
9. He, K., Zhang, X., Ren, S., Sun, J.: Deep residual learning for image recognition. In: 2016 IEEE Conference on Computer Vision and Pattern Recognition (CVPR), pp. 770–778. IEEE, Las Vegas, NV, USA (2016)
10. Simonyan, K., Zisserman, A.: Very deep convolutional networks for large-scale image recognition. arXiv:1409.1556 [cs]. (2015)
11. Iandola, F.N., Han, S., Moskewicz, M.W., Ashraf, K., Dally, W.J., Keutzer, K.: SqueezeNet: AlexNet-level accuracy with 50x fewer parameters and $\leq$ 0.5MB model size. arXiv:1602.07360 [cs]. (2016)
12. Szegedy, C., et al.: Going deeper with convolutions. In: 2015 IEEE Conference on Computer Vision and Pattern Recognition (CVPR), pp. 1–9. IEEE, Boston, MA, USA (2015)
13. AIKO-KLHK Application Android smarthphone (2021). https://play.google.com/store/apps/details?id=id.codepresso.woodid&hl=th&gl=US. Accessed 10 Aug 2021
14. The World Bank (2020). Forest Area (% of land area), https://data.worldbank.org/indicator/AG.LND.FRST.ZS?locations=TH. Accessed 07 Jan 2022

15. World Development Indicators Database (2020). https://databank.worldbank.org/views/reports/reportwidget.aspx?Report_Name=CountryProfile&Id=b450fd57&tbar=y&dd=y&inf=n&zm=n&country=THA. Accessed 07 Jan 2022
16. The Forest Research and Development Office. General Characteristics of Thai Timbers, pp. 1–3. S. Mongkol Press LTD., PART (2020)
17. Double Helix Tracking Pte Ltd. (2022) Services. https://www.doublehelixtracking.com. Accessed 07 Jan 2022

Combining Logistic Regression Analysis with Data Mining Techniques to Predict Diabetes

Ratchaneewan Paisanwarakiat[1], Anamai Na-udom[1(✉)], and Jaratsri Rungrattanaubol[2]

[1] Department of Mathematics, Naresuan University, Phitsanulok 65000, Thailand
{ratchaneewanpa64,anamain}@nu.ac.th
[2] Department of Computer Science and Information Technology, Naresuan University, Phitsanulok 65000, Thailand
jaratsrir@nu.ac.th

Abstract. This research aims to construct the predictive model for diabetes using a real dataset from the hospital. We propose an idea of combining the logistic regression analysis with some data mining techniques, including a decision tree with algorithms J48 and LMT, Naïve Bayes and Artificial Neural Network (ANN), to construct the predictive models. The advantage of logistic regression analysis is that it can identify variables affecting diabetes with four selection methods. The performance of the predictive models is evaluated by the accuracy, sensitivity, and specification on the test data, based on 5-fold cross validation method. The results show that the decision tree with LMT outperforms with the highest accuracy, while ANN has the highest sensitivity, and the logistic regression has the highest specificity. The study has also shown that by combing the logistic regression analysis, the accuracy of J48 and Naïve Bayes increased by 1–2%, while the accuracy of LMT decreased and ANN increased in some cases. The relevant variables affecting the diabetes are Systolic Blood Pressure, Diastolic Blood Pressure, Heart Rate and Fasting Blood Sugar, respectively.

Keywords: Diabetes · Logistic regression · Decision tree · J48 · LMT · Artificial Neural Network · Naïve Bayes

1 Introduction

Diabetes Mellitus is a serious, long-term, or chronic condition that occurs when there are raised levels of glucose in a person's blood, because their body cannot produce any or enough of the hormone insulin or cannot effectively use the produced insulin. Diabetes can be divided into three types: Type 1 Diabetes Mellitus, Type 2 Diabetes Mellitus and Type 3 Gestational Diabetes Mellitus. Moreover, Diabetes can lead to many severe illnesses such as blindness, kidney failure, heart attacks and stroke, and probably cause of death.

In 2019, International Diabetes Federation (IDF) reported that there are approximately 463 million individuals living with diabetes globally and predicted that the number of Diabetes cases will rise to 573 million by 2030 and 700 million by 2045

P. Meesad et al. (Eds.): IC2IT 2022, LNNS 453, pp. 88–98, 2022.
https://doi.org/10.1007/978-3-030-99948-3_9

[1]. With this large number of cases and growing tendency, many countries have tried to tackle the problems by using various methods such as analyzing the key factors affecting Diabetes, giving a guideline for prevention including nutrition, exercise, and daily activities. Hence, many researchers across the world have attempted to analyze the dataset or records of Diabetes patients to extract the factors related to Diabetes and construct the predictive models using both statistical methods and data mining techniques such as in Canada and India [2–4]. Most of the predictive models aim to classify the result as 'yes' and 'no' for Diabetes, or to classify the Type 1, Type 2 and Type 3 of Diabetes. Many of them have used the logistic regression model, which is one of the statistical modeling techniques, in a comparison with a variety of data mining classification techniques such as Decision Tree, Naïve Bayes, K-Nearest Neighbor, Support Vector Machine and Artificial Neural Network [5–9]. The real dataset was used regarding various factors or attributes related the local studying areas. Most of the dataset are usually unbalanced. It contains the number of records saying 'non-diabetic' a lot more than 'diabetic'. This is one important issue to consider and take care when constructing the model.

In this paper, we attempt to combine the logistic regression analysis with the data mining techniques to get an alternative method for the predictive model construction by using the real dataset on diabetes patients in a hospital in Thailand. It was reported that in 2018 there were 2,659,940 diabetes patients and most of them lived in the Northern part of Thailand [10]. Hence, it would be useful to extract the relevant factors of diabetes patients and obtain the accurate predictive model based on the recorded data.

The objective of this paper is to apply the logistic regression to analyze factors affecting diabetes and then construct the predictive models using the logistic regression and data mining techniques including Decision Tree using the J48 and LMT algorithm Naïve Bayes and Artificial Neural Network. and compare their performance. We focus on the most accurate predictive models, how the logistic regression analysis can improve the accuracy of data mining technique models and the relevant factors of diabetes.

2 Logistic Regression

Logistic Regression has been widely used to study the relationship between the probability of an event of interest and the probability of that event occurring. If there are only two values of dependent variable or output (i.e., 'yes' or 'no'), it is called as a "binary logistic regression". Whereas the independent variables or input attributes can be quantitative or qualitative or a combination of them. Binary logistic regression model can be expressed as an Eq. (1). The link must be linearized in regression analysis. It is defined as 'odds ratio' or 'logit', which is a linear relationship that can be expressed as an Eq. (2) [11].

$$P(Y=1)=\frac{e^{\beta_0+\beta_1X_1+\cdots+\beta_pX_p}}{1+e^{\beta_0+\beta_1X_1+\cdots+\beta_pX_p}} \tag{1}$$

$$log\left(\frac{Z}{1-Z}\right)=logit(z)=\beta_0+\beta_1X_1+\cdots+\beta_pX_p \tag{2}$$

$$Z=\frac{e^{\beta_0+\beta_1X_1+\cdots+\beta_pX_p}}{1+e^{\beta_0+\beta_1X_1+\cdots+\beta_pX_p}} \tag{3}$$

where, Z is a probability that an event occurs.

X_i is the independent variables.

β_i is the regression coefficients.

The logistic regression model is constructed by selecting input variables, which are significantly related to the output. The process of variable selection is started by searching for the statistically significant input variables, in which these variables can be performed together to obtain the accurate predictive model. The method of selecting variables can be classified into four categories: Basic, Forward Selection, Backward Elimination and Stepwise Regression.

Basic: A procedure for variable selection variables is done by a given level of significance for selection.

Forward Selection: A procedure begins with assuming that there are no input variables in the model other than the intercept. Then try to find an optimal subset of variables by inserting variables into the model one at a time. The variable selected to include in the equation is the one that has the largest simple correlation with the output. The process inserts variables until obtaining a suitable model.

Backward Elimination: By contrast with Forward selection, Backward elimination attempts to find a good model by working in the opposite way. Since it starts by removing variables from the model until a suitable model is met.

Stepwise Regression: Forward and Backward procedures give various possible combinations of variables. The stepwise regression algorithm is a modification of forward selection, in which at each step when input variables added into the model are tested with the previous by partial F (or t) test statistic. Since a new variable added may be redundant with the variables i.e., same variables are probably added in the model. The process of dropping that variable is decided based on the partial F (or t) test statistic [11].

3 Data Mining Techniques

3.1 Decision Trees

Decision trees (DT) are one of the popular classification techniques in data mining. The dataset is analyzed under the principle of the tree-structure concept, which comprises of a root node, internal nodes, branches, and leaves. A process of building a tree is based on selecting variables to be a root or nodes by measuring the impurity of the variables against the output or class label, then decide branching with respect to the selected node as depicted in Fig. 3 [12]. Each variable value is normally measured by the information and gain as given in (4) and (5).

$$Info(D) = -\sum_{i=1}^{m} p_i log(p_i) \tag{4}$$

where, D is a dataset of interest and m is a number of groups of output values.

p_i is a probability that an arbitrary tuple in D based on a current interest variable (Fig. 1).

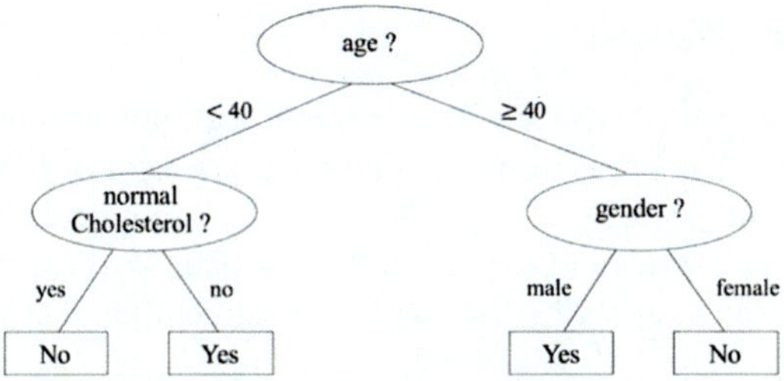

Fig. 1. An example of decision tree model.

The information gain for each variable is extended from (4) as in Eq. (5).

$$Gain(A) = -\sum_{i=1}^{k} \frac{n_i}{n} info(i) \tag{5}$$

where, A is a variable and k is a number of groups of variable values.

We used two DT algorithms which are J48 or C4.5 and LMT. As J48 is one of the most popular DTs and LMT is a DT with a combination of regression equations.

J48 or C4.5 Decision Tree Algorithm: It is a modified decision tree by extending the computation of the information gain of each variable as shown in Eq. (6) and (7). The variable with the highest value of *GainRatio* is selected to be a root or node.

$$SplitInfo(A) = \sum_{j=1}^{v} \frac{|D_j|}{|D|} \times log_2\left(\frac{|D_j|}{|D|}\right) \tag{6}$$

$$GainRatio(A) = \frac{Gain(A)}{SplitInfo(A)} \tag{7}$$

Logistic Model Tree Algorithm (LMT): A root or node of the tree is selected based on the logistic regression and J48 or C4.5 algorithmic [13].

3.2 Naïve Bayes

The Naïve Bayes classifier is a simple algorithm based on a probabilistic classifier derived from Bayes' theorem. The maximize $P(c|x)$ is the predicted output or class and is defined in Eq. (8).

$$P(c|x) = \frac{P(x|c)P(c)}{P(x)} = P(x_1|c)P(x_2|c)...P(x_n|c) \tag{8}$$

where, $P(c|x)$ is the posterior probability of the class given the predictors.
$P(c)$ is the prior probability of the class.
$P(c|x)$ is the likelihood which is the probability of the predictor given the class.
$P(x)$ is the prior probability of predictor.

3.3 Artificial Neural Network

Artificial Neural Network (ANN) is an algorithmic learning method developed based on a human brain, which consists of many complex systems of cells. A single system or node is structed as depicted in Fig. 2 [14]. The inputs and pre-calculated weights produce the predicted output as defined in Eq. (9). An ANN model is constructed by connecting a set of nodes into a layer, applying the weight adjustment process as the key learning algorithm.

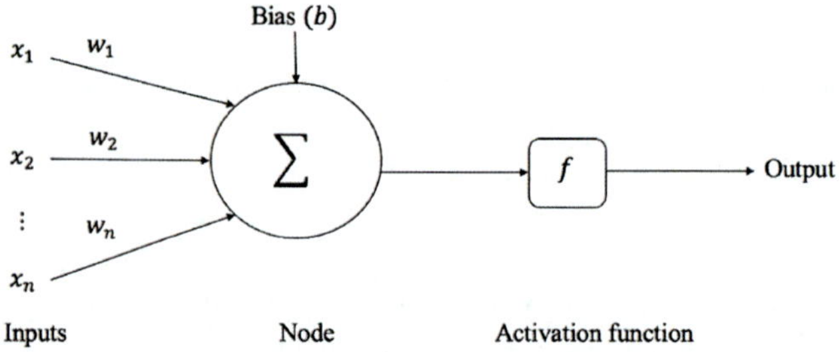

Fig. 2. A structure of a basic node in Artificial Neural Network.

$$y = f\left(\sum_{i=1}^{n} x_i w_i + b\right) \tag{9}$$

where, f is an activation function, which is a sigmoid function and $f(x) = \frac{1}{1+e^{-x}}$.
x_i are the independent variables.
w_i is a weight for each connection, and b is a bias at the node.

4 Methodology

The main goal of this research is to construct the predictive models for diabetes by combing logistic regression analysis with data mining techniques. The process begins with data understanding and preprocessing, the logistic regression predictive modeling, the data mining predictive modeling and the model performance evaluation.

4.1 Data Understanding and Preprocessing

The dataset collected from Information Unit at Naresuan University Hospital during January 1–June 30, 2020, contains thirteen input variables and one output with two values ('non-diabetic' and 'diabetic'). There are a total of 5,081 records with 3,934 records for 'non-diabetic' and 1,147 for 'diabetic'). The dataset is unbalanced since there are about 77% of the dataset for 'non-diabetic'. As displayed in Table 1, the average of Age, Body Weight and Height are 59.94, 65.38 and 159.93, respectively. The dataset contains an older group of patients and the same amount of male and female.

As mentioned previously, when dealing with an unbalanced dataset the process of designing a training data and a test data is very important. A training set is used for the model construction, while a test data is for the model evaluation. In this research, we used k-fold cross validation in constructing the predictive models and evaluating. The k-fold concept is randomly partitioning a dataset into k groups, reserve one group as a test set and the others as a training set.

Table 1. Description, unit, type and statistics values of input variables (X) and output (Y).

Variable	Description	Unit	Type	Percent, Average ± SD
X1	Gender	0 = male, 1 = female	Category	2,177: 2,904 (42.85%: 57.15%)
X2	Age	year	Numeric	59.94 (±13.85)
X3	Systolic Blood Pressure	mmHg	Numeric	135.47 (±19.19)
X4	Diastolic Blood Pressure	mmHg	Numeric	74.05 (±12.47)
X5	Heart Rate	mmHg	Numeric	78.81 (±12.99)
X6	Body Weight	kg	Numeric	65.38 (±14.31)
X7	Height	cm	Numeric	159.93 (±8.71)
X8	Body Mass Index	kg/m^2	Numeric	25.52 (±5.10)
X9	Fasting Blood Sugar	mg/dL	Numeric	114.78 (±36.06)
X10	Cholesterol	mg/dL	Numeric	178.18 (±37.63)
X11	Triglyceride	mg/dL	Numeric	134.83 (±63.88)
X12	HDL	mg/dL	Numeric	52.62 (±14.07)
X13	LDL	mg/dL	Numeric	98.60 (±33.59)
Y	**Output**	**0 = non-diabetic 1 = diabetic**	**Category**	**3,934:1,147 (77.43%:22.57%)**

As displayed in Fig. 3, we set k to 5 and partitioned the dataset into 5 set of data by ensuring that the ratio of ‘non-diabetic’ and ‘diabetic’ remains the same. In Fig. 3, first we partitioned each set of the data ‘diabetic’ and ‘non-diabetic’ into 5 groups, then merged each of them to get the set of partition (i.e., Data1, Data2, …, Data5). With 5-fold method, we repeat constructing the predictive model five times and calculate the average accuracy, which is defined as Eq. (10).

$$Accuracy = \frac{1}{5}\sum\nolimits_{i=1}^{5} Accuracy\_test_i \tag{10}$$

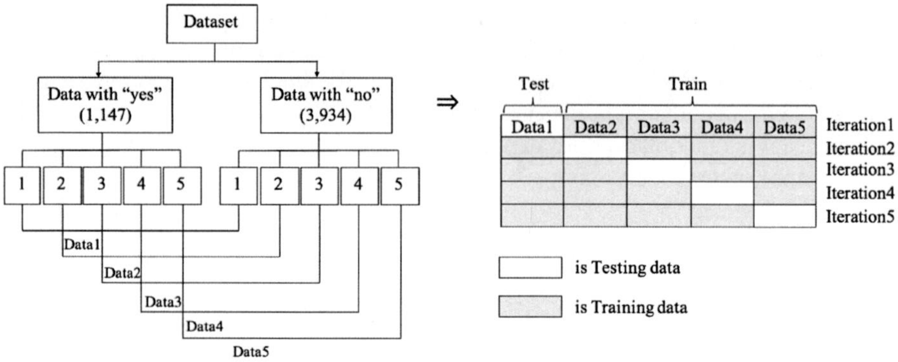

Fig. 3. A process of 5-fold cross validation applying to the dataset.

4.2 The Logistic Regression Predictive Modeling

We wrote R-program to construct the logistic regression model, using four variable selection methods as mentioned in Sect. 2. The statistical significance level is 0.05 for all methods. The logistic regression equation and variables selected for each method is shown as follows.

Basic: Systolic Blood Pressure, Diastolic Blood Pressure, Heart Rate, Body Weight, Height and Fasting Blood Sugar.

$$P(Y=1)=\frac{1}{1+e^{-(\beta_0+\beta_3X_3+\beta_4X_4+\beta_5X_5+\beta_6X_6+\beta_7X_7+\beta_9X_9)}} \quad (11)$$

Forward Selection: Systolic Blood Pressure, Diastolic Blood Pressure, Heart Rate, Body Mass Index, Fasting Blood Sugar and Cholesterol.

$$P(Y=1)=\frac{1}{1+e^{-(\beta_0+\beta_3X_3+\beta_4X_4+\beta_5X_5+\beta_8X_8+\beta_9X_9+\beta_{10}X_{10})}} \quad (12)$$

Backward Elimination: Systolic Blood Pressure, Diastolic Blood Pressure, Heart Rate, Body Weight, Height, Fasting Blood Sugar, HDL and LDL.

$$P(Y=1)=\frac{1}{1+e^{-(\beta_0+\beta_3X_3+\beta_4X_4+\beta_5X_5+\beta_6X_6+\beta_7X_7+\beta_9X_9+\beta_{12}X_{12}+\beta_{13}X_{13})}} \quad (13)$$

Stepwise Regression: Systolic Blood Pressure, Diastolic Blood Pressure, Heart Rate, Body Mass Index, Fasting Blood Sugar and Cholesterol.

$$P(Y=1)=\frac{1}{1+e^{-(\beta_0+\beta_3X_3+\beta_4X_4+\beta_5X_5+\beta_8X_8+\beta_9X_9+\beta_{10}X_{10})}} \quad (14)$$

4.3 The Data Mining Technique Predictive Modeling

When constructing a predictive model with data mining techniques, we usually include all variables into a model, without examining a level of relationship of variables with

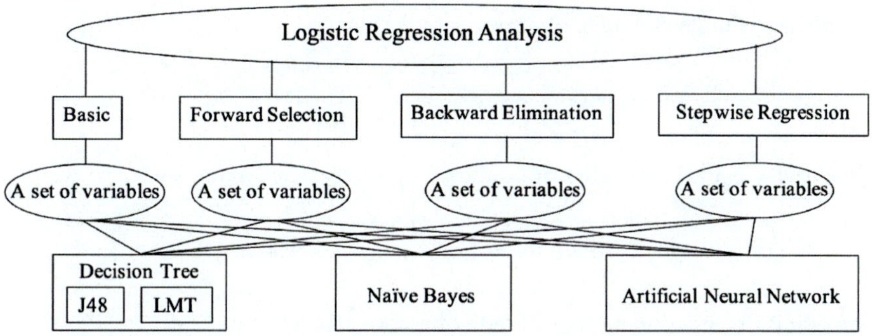

Fig. 4. Combining a logistic regression (four variable selections) with data mining techniques.

the output. Although, some researchers may try-and-evaluate by adding and deleting variables manually with or without statistical background.

In this research, we combined a set of variables selected by the logistic regression analysis as the input variables to the data mining technique modeling, which are DT with algorithm J48 and LMT, Naïve Bayes and ANN as shown in Fig. 4. We extended the R code from a previous section to construct the models for each technique, using 5-fold. Then we repeated constructing each modeling technique (with a set of variables) five times and calculated the average of each performance evaluation value.

The parameters of J48, LMT and Naïve Bayes are set as the default value. While the parameters of ANN are experimented for the best performance, which are a number of nodes varying from 6 to 10 nodes, a momentum and learning rate varying 0.1, 0.2 and 0.3, and the epoch is set to 100.

4.4 The Model Performance Evaluation

The performance of the predictive model is measured by these three criteria: Accuracy, Sensitivity and Specificity. Each criterion is calculated in Eq. (15)–(17) (Table 2).

Table 2. A Confusion Metric displaying the result of a predictor against the actual.

Actual values	Predicted values	
	Positive	Negative
Positive	*TP*	*FN*
Negative	*FP*	*TN*

Where, *TP* is a number of predicting Positive correctly.
TN is a number of predicting Negative correctly.
FP is a number of predicting Negative as Positive.
FN is a number of predicting Positive as Negative.

Accuracy is the percentage of correct predictions.

$$Accuracy = \left(\frac{TP + TN}{TP + FP + FN + TN}\right) \times 100 \tag{15}$$

Sensitivity is the probability of predicted correctly when a disease.

$$Sensitivity = \left(\frac{TP}{TP + FN}\right) \times 100 \tag{16}$$

Specificity is the probability of predicting correctly when not a disease.

$$Specificity = \left(\frac{TN}{TN + FP}\right) \times 100 \tag{17}$$

5 Results

The performance evaluation values for each predictive model are calculated on the test data equation. The 5-fold cross validation gives five results for each value, therefore, the values presented in Table 3 are the average among 5-folds. The 'Full' means using all variables in the model.

Table 3. The average performance evaluation values of each predictive model.

		Full	Basic	Forward	Backward	Stepwise
Logistic regression	Accuracy		***80.6926***	80.5943	80.5943	80.5943
	Sensitivity		31.9077	31.2106	**31.9966**	31.2106
	Specificity		94.9163	**94.9925**	94.7639	94.9925
J48	Accuracy	81.0074	81.8933	81.8736	***81.9132***	81.8736
	Sensitivity	54.0558	55.0165	56.0611	**56.6698**	56.0611
	Specificity	88.8661	88.6121	**89.4001**	89.2733	**89.4001**
LMT	Accuracy	***83.1333***	82.3462	82.1884	82.3462	82.1884
	Sensitivity	56.2324	56.9337	**58.8464**	57.1050	58.8464
	Specificity	**90.9762**	89.7562	88.9936	89.7054	88.9936
Naïve Bayes	Accuracy	80.0826	***81.5390***	81.3618	81.1254	81.3618
	Sensitivity	56.0598	**60.7349**	59.8732	52.6546	59.2690
	Specificity	**86.9492**	86.5045	86.7963	86.6306	86.7963
ANN (8, 0.2, 0.3)	Accuracy	81.5205	81.1270	***81.6192***	81.1023	***81.6192***
	Sensitivity	62.6344	61.1620	**63.0726**	62.8753	**63.0726**
	Specificity	85.0598	85.2185	**85.5298**	84.0920	**85.5298**

The optimal parameters for ANN are 1) a number of nodes is 8, 2) momentum rate is 0.2 and, 3) learning rate is 0.3. According to the accuracy, LMT with full performs the best, followed by J48 with backward and ANN with forward and stepwise. When considering the sensitivity, ANN with forward and stepwise performs the best, followed by Naïve Bayes with basic. With the specificity, logistic regression with forward performs best. For interpretation, LMT outperforms in giving the correct prediction, while ANN gives the highest probability in predicting 'diabetic' and Logistic Regression predict 'non-diabetic' with the highest probability.

Table 4. The percentage of the accuracy differences from using all variables (Full).

	J48	LMT	Naïve Bayes	ANN (8,0.2,0.3)
Full	0.0000	0.0000	0.0000	0.0000
Normal	1.0936	−0.9468	1.8187	−0.4828
Forward	1.0693	−1.1366	1.5973	0.1210
Backward	1.1182	−0.9468	1.3021	−0.5131
Stepwise	1.0693	−1.1366	1.5973	0.1210

When considering the performance accuracy of each data mining technique after applying a logistic regression analysis, the results in Table 4 indicates that J48 and Naïve Bayes performs better around 1–1.8%, while LMT accuracy reduces around 1–1.1%, and ANN performs lightly better with forward and stepwise.

6 Conclusion

The objective of the research is to construct the predictive model for diabetes using the idea of combining logistic regression and data mining techniques. According to the result presented in the previous section, it can be concluded that LMT is the most accurate model, however its sensitivity is less than those of ANN models. The result obtained from variable selection methods using logistic regression analysis indicates that the significant variables affecting diabetes are Systolic Blood Pressure, Diastolic Blood Pressure, Heart Rate and Fasting Blood Sugar. Finally, by combining or applying the logistic regression analysis to the data mining technique, the predictive accuracy of J48 and Naïve Bayes increases.

Acknowledgement. This research is supported by Thailand Science Research and Innovation (TSRI) grant contract no. R2564B006.

References

1. IDF: IDF Diabetes Atlas, 9th edn. https://www.diabetesatlas.org. Accessed 16 May 2020

2. Selvakumar, S., Kannan, K.S., GothaiNachiyar, S.: Prediction of diabetes diagnosis using classification based data mining techniques. Int. J. Stat. Syst. **12**(2), 183–188 (2017)
3. Perveen, S., Shahbaz, M., Guergachi, A., Keshavjee, K.: Performance analysis of data mining classification techniques to predict diabetes. Procedia Comput. Sci. **82**, 115–121 (2016)
4. Sisodia, D., Sisodia, D.S.: Prediction of diabetes using classification algorithms. Procedia Comput. Sci. **132**, 1578–1585 (2018)
5. Kumar, P.S., Umatejaswi, V.: Diagnosing diabetes using data mining techniques. Int. J. Sci. Res. Pub. **7**(6), 705–709 (2017)
6. Manimaran, R., Vanitha, M.: Prediction of diabetes disease using classification data mining techniques. Int. J. Eng. Technol. **9**(5), 3610–3614 (2017)
7. Sinsomboonthong, S.: An efficiency comparison of diabetes prediction. Sci. Technol. Asia **26**(2), 195–207 (2018)
8. Kumar, C., Singh, N., Singh, J.: Prediction of diabetes using data mining algorithm. Int. J. Res. Appl. Sci. Eng. Technol. **7**(2), 773–780 (2019)
9. Sunge, A.S., Warnar, H.L., Heryadi, Y., Abdurachman, E., Soewito, B., Gaol, F.L.: Prediction diabetes mellitus using decision tree models. In: 2019 International Congress on Applied Information Technology, pp. 1–6. IEEE, Indonesia (2019)
10. AESR: https://apps-oe.moph.go.th/boeeng/download/AW_Annual_Mix%206212_14_r1.pdf, last accessed 2020/10/16
11. Montgomery, D.C., Peck, E.A., Vining, G.G.: Introduction to Linear Regression Analysis, 5th edn. John Wiley & Sons Inc., Hoboken (2012)
12. Han, J., Kamber, M., Pei, J.: Data Mining: Concepts and Techniques, 3rd edn. Morgan Kaufmann, San Francisco (2012)
13. Landwehr, N., Hall, M., Frank, E.: Logistic model trees. Mach. Learn. **95**(1–2), 161–205 (2005)
14. Larose, D.T., Larose, C.D.: Discovering Knowledge in Data: An Introduction to Data Mining, 2nd edn. John Wileys & Sons, New York (2014)

AutoRat: Automated Risk Assessment Tool for Network Mapper Scanning

Atit Sethapanee, Tanakorn Nimitrchai, and Somchart Fugkeaw(✉)

School of Information, Computer, and Communication Technology, Sirindhorn International Institute of Technology, Thammasat University, Pathum Thani 12120, Thailand
{6122780271,somchart}@siit.tu.ac.th

Abstract. Network vulnerability discovery can generally be done using vulnerability scanning tools that are able to identify possible threats or flaws of network services, ports, devices, protocols, operating systems, and software connected to a network. Most existing tools give a list of vulnerabilities on network-related services and they are used in another stage of the analysis such as risk assessment, penetration test, and re-programming. However, the automatic compilation of the detected vulnerabilities into the risk assessment is crucial for timely risk mitigation and optimized penetration test process. This paper proposes a design and implementation of a risk assessment tool called Auto-RAT which delivers an automatic risk assessment from Nmap, a network enumeration tool supporting the penetration test process. To enable the usability and immediate risk analysis of the Nmap, Auto-RAT is designed to realize the process of Nmap and its result for the automatic compilation into the risk assessment engine. We propose a set of algorithms to convert Nmap scan results into risk assessment model where the risk scoring is calculated using the Common Vulnerability Scoring System (CVSS). Finally, the system generates final risk assessment score and possible risk mitigation strategies for the system evaluation. We run the simulation by comparing the overlapped results between our AutoRAT and Nessus scan result to demonstrate that AutoRAT can effectively reduce the set of vulnerabilities that will be executed in the penetration test process.

Keywords: Risk assessment · Penetration test · Nmap · Vulnerability scoring system

1 Introduction

Vulnerability scanning (VA) and penetration testing (PenTest) are essential for network and system security management since they can reveal an abnormality/error in networking, access control, and application. These activities require multiple steps for carrying out the assessment and test. Specifically, 'Pentest tool' is a term used to describe a software application that functions as a cyber security threat checking tool. It validates the detected vulnerabilities whether they are genuine and may provide potential impact to the business. In essence, network enumeration is a prerequisite task done before Pentest tools are employed to scan the vulnerabilities and exploit the payload to test the security

P. Meesad et al. (Eds.): IC2IT 2022, LNNS 453, pp. 99–110, 2022.
https://doi.org/10.1007/978-3-030-99948-3_10

threats. Some VA and PenTest tools provide visualization features for efficient analysis [1, 2].

In the enumeration phase, the pentester can run the tool such as Nmap [3] which is widely used for testing and vulnerabilities in a network system. The main usage of Nmap is port scanning, version detection, OS detection, and it serves as the reconnaissance for a pen-testing process. There are many underlying features which Nmap is capable of, including providing quick, detailed scans of network traffic, showing ports status, as well as displaying the OS they are running on. Therefore, the results obtained from running Nmap are very useful for pre-analysis and assessment of possible risks or vulnerabilities that might exist in the network system.

Currently, there are no pen-testing tools that allow for automatic risk assessment by realizing the Nmap result as the first-tier risk consideration. Typically, most organizations perform risk assessments after the Pentest is done. This is because some VA or PenTest tools provide risk assessment features for those vulnerabilities detected. However, pentest conducted in a large-scale system environment such as those with many hosts, URLs take several days or weeks to complete. Assessing risk from a pen-testing result will require a considerable amount of knowledge in using tools and exploitable vulnerabilities. Therefore, it would be highly beneficial if a sound output of the risk assessment can be generated in the enumeration stage.

To the best of our knowledge, there are no existing tools that provide the automatic conversion of the Nmap results into the risk assessment as a part of the Pentest process. Analyzing the Nmap scan through the report could lead to some vulnerability to be overlooked or handled inappropriately. Automation of the assessment process will enable an immediate response to pen-testing results.

In this paper, we propose a design and implementation of Auto-RAT, an automated security assessment tool based on the Nmap result. Our proposed system is designed to be able to determine the vulnerability's exploits or risks from the Nmap scan results. All detected vulnerabilities are systematically checked through Vulner's common vulnerabilities and exposures (CVE) database [4] and they are classified by their attack vector and threat level for ease of analysis and risk assessment. Then, the classified vulnerabilities are fed into our risk assessment model and the outputs of their prioritized risk score with associated risk mitigation strategy are dynamically generated.

2 Related Work

Nmap has been recognized as an effective network vulnerabilities discovery tool. Several research works employ Nmap [5] for scanning the vulnerabilities and the scan results are then analyzed. In [6] the authors presented a performance evaluation of network vulnerability scanning tools including Nessus, OpenVAS, and Nmap Scripting Engine performance and reported the results on how they can be used to automate the process of risk assessment in an organization.

In [7], Shah et al. tackled the network traffic analysis problem during the reconnaissance phase of the penetration test by using Nmap scan results for network traffic auditing and determining large volumes of hosts. Then, the optimization strategy for network traffic and the time of the specific task is applied.

Recently, Zhang et al. [8] conducted an experiment to show the capabilities of two scanners, Nmap and Masscan, and discuss the applications of these two scanners. Nmap is proven as real-time, and its result has potential for diagnosis.

In [9], Rohrmann et al. proposed a parallelized Nmap scan to support an effective way to anonymously collect internet scan data. However, all the above works have not emphasized converting Nmap results into the risk assessment and evaluation for reducing the penetration test effort. To analyze the presence and correlation of the network vulnerabilities, several research works proposed graph-based modeling schemes [10–17] as an effective way for risk and vulnerability management.

In [10], the authors proposed A probabilistic directed graph model namely Hidden Markov Model (HMM), which is used to construct a probabilistic mapping between network observation and attack status. The model is able to model multiple attack steps in the nodes and edges of an attack graph model where the probability value is calculated and assigned. However, this scheme did not focus on risk assessment.

In [11], the authors applied the Bayesian network to model the network attack and proposed the Bayes formula to calculate the probability of a new node probability calculation formula and attack paths. The proposed model can support quantitative network vulnerability assessment. The graph model is generated based on CVE stored.

In [12], Wang et al. proposed a graph-based model to discover the optimal attack path by considering the maximum flow. The common vulnerability scoring system was proposed to calculate the risk which is determined by the attack path quantification degree. However, the authors do not provide solutions to mitigate the risk.

Nevertheless, all of the above research works have not entailed the minimization of risk during the vulnerability assessment which is a part of the pentest process. Also, Nmap scan results have not been analytically used for risk assessment and risk mitigation plan generation by any works.

3 Our Proposed System

3.1 System Overview

This section presents the framework of the AutoRAT and a detailed description of its components. Figure 1 exhibits the system model of AutoRAT.

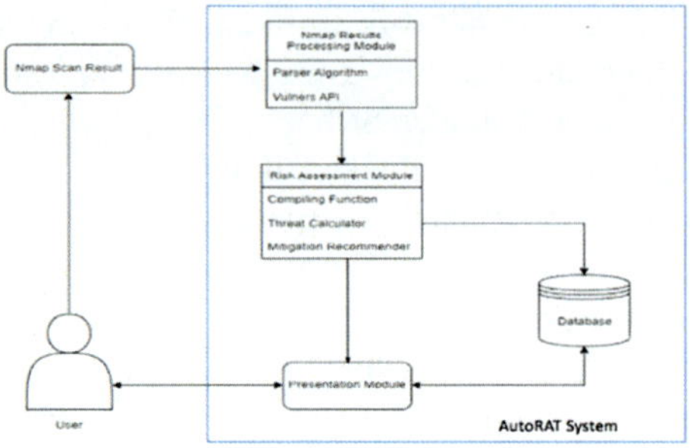

Fig. 1. AutoRAT system model.

The system model consists of the following entities.

1) User can be either a system administrator or pentester who is responsible for running the AutoRAT system, obtaining the scan results from Nmap.
2) Nmap scan result is a set of scan reports generated from the Nmap tool.
3) AutoRAT system consists of four main components, including Nmap result processing module, risk assessment module, presentation module, and a database used to store the resulting data and logs.

The next sections give more details of the first three core modules of our AutoRAT.

3.2 Nmap Result Processing Module

This module is responsible for processing the Nmap scan result. It allows the user to interact with the GUI and import an output of a Nmap scan. To render the efficient analysis of the vulnerabilities detected as well as to reduce the number of vulnerabilities to be tested in the real pentest process where test exploitation is required, we normalize the scan results before carrying out the pentest operation. Here, we devise a parser algorithm to separate the CVEs information including the port, services, and CVE ID. The parser will identify the vulnerabilities found in the scan and forward the data of a fixed format to the Vulners API. In our design, if the vulnerabilities exist in Metasploit which is a penetration test tool, the parser will automatically insert them into a temporary dictionary and add an array with all the CVE ID for immediate exploitation. The parsing process is carried out with the help of the python regular expression library. The pseudocode below shows the algorithmic process of the parser module. The system initializes a dictionary called the CVEdictionary to contain all the data parsed from the inputted file. Additionally, an array namely the CVEarray will contain all the CVE IDs found throughout the scan result.

Parser Algorithm

```
Begin
  Input nmap Scan Result
  CVEdictionary = {}
  CVEarray = [ ]
  For every line in input:
    if line length != 0 Then
      if first char in line is not a digit Then
        cleanline = replace consecutive spaces with single space
        linesplit = split the line by space
        port = first split
        state = second split
        service = third split
        if fourth split exists Then
          version = fourth split
        else
          version = False
      else if first char in line == '|' Then
        metasploitavailable = False
        cleanline = replace consecutive spaces with single space
        cve = find "CVE..." in line
        if cve exists Then
          cveData = split cve by space
          if last char of cveData == "/" Then
            cveID = cvedata sliced last char
            metasploitCommand = find string "MSF:"
          else
            cveID = first split of cveData
            cveScore = second split of cveData
            append cveID to CVEarray
            create an entry of the CVEID in CVEdictionary
            append port, state, service to the entry
            if version exists Then
              append version to the entry
            if metasploitavailable exists Then
              append metasploitCommand to the entry
  Output CVEarray, CVEdictionary
End
```

To realize the characteristics of each vulnerability detected for its usage in the risk assessment process, its CVE ID will then be used to query the Vulners DB [18]. Then, the documentation providing the details of each CVE is obtained. Figure 2 illustrates the flow of inputted CVE ID into AutoRAT and its states transitions. Note that, we only consider the vulnerability found in VulnersDB, but the possibility of integrating more databases is possible in the future.

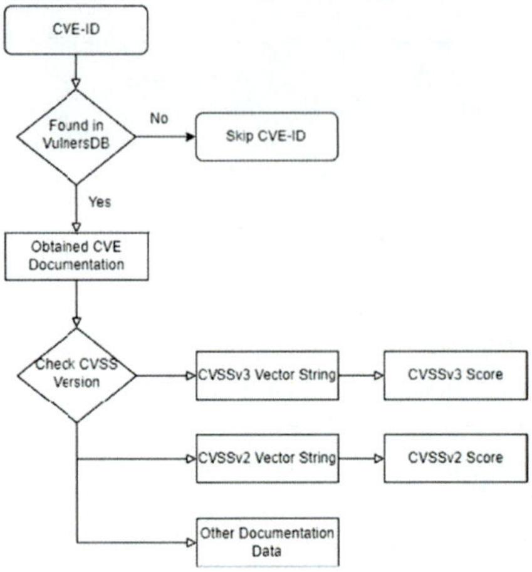

Fig. 2. Correlation of the CVE IDs.

With the documentation, we can refer to the required variables to calculate the threat score using the CVSS standard scoring system [19, 20] for calculating the risk. These variables include the attack complexity and the impact on the confidentiality, integrity, availability (CIA), the pre-calculated CVSS scores, and other related traits of the given vulnerability.

The CVSS system consists of three metric groups, the base score, temporal score, and environmental score. The base score reflects the general characteristics of a vulnerability, the temporal score reflects the vulnerability score including the factors which are temporal, and lastly, the environmental metric includes the factor of the computing environment such as the mitigations currently available. The CVSS scoring system uses multiple variables to calculate for each metric as shown in Fig. 3. Additionally, the CVSS scoring system also uses a standard vector that describes the metrics called the vector string which will be utilized by AutoRAT. AutoRAT can precisely utilize these vector strings for any further computation required by the compilation in the Risk assessment module.

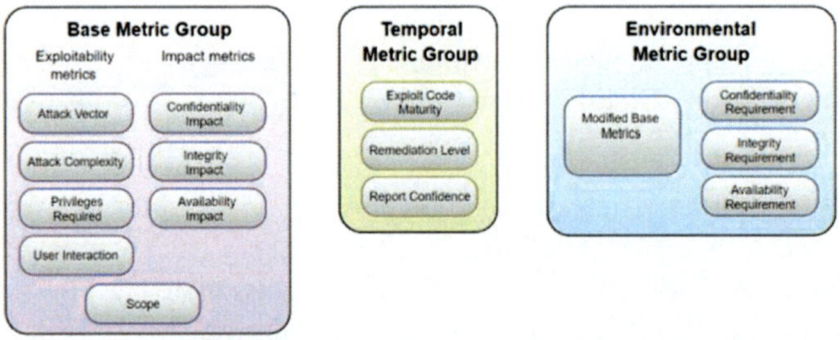

Fig. 3. CVSSv3 metric groups [19].

3.3 The Risk Assessment Module

In the Risk Assessment module, we propose a CVE compiler as our risk assessment engine based on the parsed Nmap result obtained from the Nmap result processing module and the documentation from the Vulners API. The model is constructed from the use of three major vulnerability analysis components: (1) VulnersDocuments dictionary containing the documentation of vulnerabilities and it is derived from the Vulners API, (2) Vulnerabilities data (CVEDictionary) which stores data parsed from the Nmap scan results, and (3) CVE ID (CVEarray) housing the list of all available CVE IDs found throughout the scan results. The two latter components are the results generated by the parser process. These variables are systematically indexed with the CVE number, which we can utilize the CVEarray to iterate through all indexes. Following the iterations, we will obtain a new dictionary called 'compileDictionary' which combines both the CVEdictionary and VulnersDocuments together to provide better accessibility and understanding of the data.

Figure 4 describes the processing of data input into the compiler and its resulting data form. To conclude, all the documentation from the Vulners API, data obtained from the Nmap scan, and the list of all CVE IDs identified in the scan are inputted into the compiler module. After the compilation, results are compiled into a single dictionary compressed with data from all the input variables.

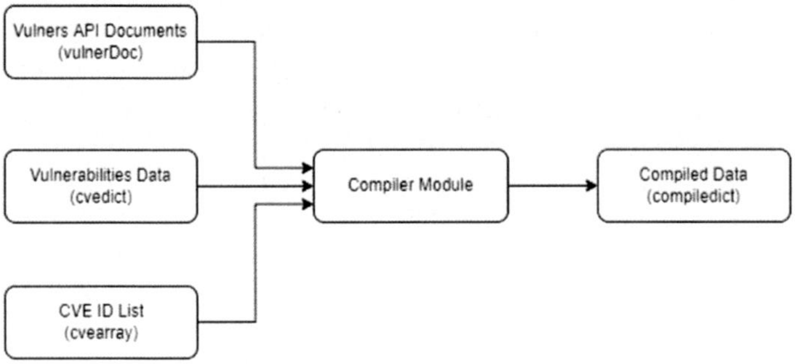

Fig. 4. The CVE compilation process.

The algorithm of the complier is shown below. The system tasks as inputs Vulners documents, CVEdictionary and CVEarray. Specifically, the system introduces a dictionary called the compileDictionary which serves as the complete compilation of all the input data.

Complier Algorithm

```
Begin
    Input VulnersDocuments, CVEdictionary, CVEarray
    compileDictionary = {}
    cvssv3metric= [array of CVSSv3 properties]
    cvssv2metric= [array of CVSSv2 properties]
    cvedataElement = ['port','state','service','version','msfcmd']
    CIAimpacts = ["integrityImpact", "availabilityImpact","confidentialityImpact"]
    for all CVE-ID in CVEarray
        add entry of CVE-ID in compileDictionary
        for each element in cvssv3metric
            if element exists in VulnersDocuments[CVEID][CVSSv3] Then
                add element to the entry
        for each element in cvssv2metric
            if element exists in Vulnersdocuments[CVEID][CVSSv2] Then
                add element to the entry
    Output compileDictionary
End
```

As detailed in the algorithm above, a set of all CVE IDs and their corresponding information compiled from all arrays inputted from the parser module will be generated. To calculate the risk of vulnerabilities detected, we use the CVSS framework and python CVSS library [20] as core schemes in which the AutoRAT invokes the CVSS function. By default, the documentation inputted to the compiler contains pre-calculated CVSS scores and its vector string. AutoRAT allows the user to alter the vector string and recalculate the CVSS scores.

Basically, each CVE number in the array is associated with the default CVSS3 or CVSS 2 score precalculated by VulnersDB. Each CVE number has its pre-configured properties, which are set in accordance with different CVSS versions. The two dominant versions used are the CVSS2 and the CVSS3, each CVE representing each threat may or may not use either or both versions. AutoRAT allows the customization of the calculation functionality to offer beneficial properties in multiple cases where the user wants to adjust the scoring system to fit their risk assessment criteria. Then, the impact on the elements of the CIA is updated. The procedure of threat calculation is presented in the pseudo-code below.

Threat Calculation Algorithm

```
Begin
  Input compileDictionary, CVEID, CVSSVersion
  allScore = {}
  if compileDictionary[CVEID][CVSSVersion] exists Then
    vectorString = compileDictionary[CVEID][CVSSVersion]['vectorString']
  if CVSSVersion is CVSSv3 Then
    Evaluation = call CVSS3 from PyPi cvss
  if CVSSVersion is CVVSSv2 Then
    Evaluation = call CVSS2 from PyPi cvss
  score = call Evaluation.scores() from PyPi cvss
  severity = call Evaluation.severities() from PyPi cvss
  add scores entries to allScore
  add severity entries to allScore
  Output allScore
End
```

The process requires the input of the compileDictionary from the compiler module, the target CVE ID, and the CVSS Version requested. The system introduces a dictionary called allScore to contain all the recomputed scores.

For each vulnerability found in the scan results, the Vulners database provides a CWE (Common Weakness Enumeration) ID indicating the specific type of vulnerability. The recommendation for each vulnerability is done by retrieving the mitigation strategy from the MITRE CWE database [21]. All these data gathered and compiled in this module will be inserted into the local application database which is utilized by the presentation module. After AutoRAT the compilation and risk assessment process, the final output is a structured set of data on all threats found.

3.4 The Presentation Module

This system module is designed to allow the users to interact with the AutoRAT with ease. Basically, the application database is a major source where this module invokes the computed risk results and remedy plan for the presentation. It consists of a dashboard showing the current statistics of the system. Basically, the compiled vulnerabilities are stored in the database and they are grouped by their profile enabling the user to perform multiple queries at once. These vulnerabilities are further classified into its sub-class as follows:

1) Network, the vulnerable component is accessible by the attackers from the network or the internet, these are considered as remote exploits.
2) Adjacent, the vulnerable component is accessible from within the same network.
3) Local, the vulnerable component is only accessible by the local system.
4) Physical, the component can be accessed by the attacker physically.

The classification of the attack vectors allows the user to manage the vulnerabilities and risk mitigation with ease. Additionally, AutoRAT further classifies the vulnerabilities by its threat scores level High, Medium, and Low, and its base scores as shown in Fig. 5.

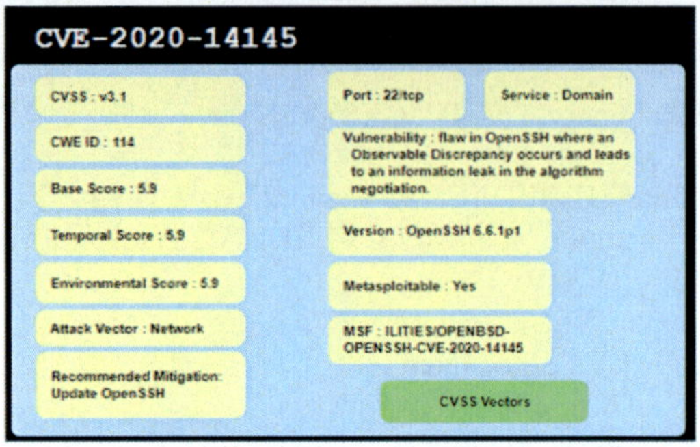

Fig. 5. Classified result.

4 Evaluation and Implementation

This section provides the analysis of the integration of AutoRAT into the Pentesting process. Firstly, in a working environment, vulnerability assessment (VA) is done before conducting a pentest. This is to acquire a set of vulnerabilities used in the pentest process. AutoRAT can assist in reducing the set of vulnerabilities required to be tested. This enables both timely mitigations of serious vulnerabilities and increases the efficiency of the pentest process.

4.1 Experiment Setting

We conducted the experiment in the simulated environment by using a virtual machine hosting a system called the Metasploitable [22]. Metasploitable is a framework created for cyber security testing which contains multiple exploitable vulnerabilities. In the experiment, we run the Nessus Essential software [23] which is a vulnerability assessment tool maintained by Tenable Inc. Simultaneously, we perform a nmap scan with the script "nmap -sV –script vulners" to scan and feed the output into the AutoRAT.

4.2 Experiment Results

After the experiment is conducted, the total number of vulnerabilities found by Nessus is 16 vulnerabilities, and 249 vulnerabilities by AutoRAT. In the scan results of Nessus, only 16 vulnerabilities are associated with a CVE-ID, therefore applicable to comparison. We can observe that there are 4 vulnerabilities found in AutoRAT which are overlapping with the Nessus results. The overlap equates to 25% of the vulnerabilities found by Nessus. Noticeably, the overlapping of the scan results and AutoRAT is subject to the target machine we conducted the test. As our test, it can confirm that the AutoRAT can effectively reduce the set of vulnerabilities that requires pentesting; this optimizes both penetration test workloads and reduces the time to address crucial vulnerabilities overlapped ones found in the network enumeration stage.

5 Conclusion

We have proposed the AutoRAT system to assess the risk of the vulnerabilities of Nmap scan results. A set of algorithms has been introduced to systematically validate and classify the results against the vulnerability database, and calculate the risk of each vulnerability. We also conducted the experiment to show that AutoRAT efficiently reduces the penetration test workload and offers timely risk mitigation. For future works, we plan to apply multi-thread programming to enhance the performance of parser and compiler algorithms.

References

1. Yuen, J., Turnbull, B., Hernandez, J.: Visual analytics for cyber red teaming. In: IEEE, Symposium on Visualization for Cyber Security (VizSec 2015), pp. 1–8 (2015). https://doi.org/10.1109/VIZSEC.2015.7312765
2. Zheng, S., et al.: PTVis: visual narrative and auxiliary decision to assist in comprehending the penetration testing process. IEEE Access **8**, 194523–194540 (2020). https://doi.org/10.1109/ACCESS.2020.3033391
3. Nmap official website. https://nmap.org/. Accessed 9 Feb 2022
4. VulDB Search. https://vuldb.com/?search. Accessed 9 Feb 2022
5. Liao, S., et al.: A Comprehensive detection approach of Nmap: principles, rules and experiments. In: Proceeding of the 2020 International Conference on Cyber-Enabled Distributed Computing and Knowledge Discovery (CyberC 2020), Chongqing, China, 29–30 October 2020, pp. 64–71 (2020). https://doi.org/10.1109/CyberC49757.2020.00020
6. Chalvatzis, I., Karras, D.A., Papademetriou, R.C.: Evaluation of security vulnerability scanners for small and medium enterprises business networks resilience towards risk assessment. In: Proceeding of IEEE International Conference on Artificial Intelligence and Computer Applications (ICAICA 2019), Dalian, China, 29–31 March 2019, pp. 52–58 (2019) https://doi.org/10.1109/ICAICA.2019.8873438
7. Shah, M., Ahmed, S., Saeed, K., Junaid, M., Khan, H., Ata-ur-rehman.: penetration testing active reconnaissance phase – optimized port scanning with Nmap tool. In: Proceeding of 2nd International Conference on Computing, Mathematics and Engineering Technologies (iCoMET 2019), Sukkur, Pakistan, 30–31 January 2019, pp. 1–6. (2019). https://doi.org/10.1109/ICOMET.2019.8673520
8. Zhang, Z., Towey, D., Ying, Z., Zhang, Y., Zhou, Z.Q.: MT4NS: metamorphic testing for network scanning. In: Proceeding of IEEE/ACM Sixth International Workshop on Metamorphic Testing (MET 2021), Madrid, Spain, 2 June 2021, pp. 17–23 (2021). https://doi.org/10.1109/MET52542.2021.00010
9. Rohrmann, R.R., Ercolani, V.J., Patton, M.W.: Large scale port scanning through tor using parallel Nmap scans to scan large portions of the IPv4 range. In: IEEE International Conference on Intelligence and Security Informatics (ISI 2017), Beijing, China, 22–24 July 2017, pp. 185–187 (2017). https://doi.org/10.1109/ISI.2017.8004906
10. Liu, S.-C., Liu, Y.: Network security risk assessment method based on HMM and attack graph model. In: Proceeding of 17th IEEE/ACIS International Conference on Software Engineering, Artificial Intelligence, Networking and Parallel/Distributed (SNPD 2016), pp. 517–522. Shanghai, China, 30–1 May/Jun 2016, https://doi.org/10.1109/SNPD
11. Zhu, X.: Computer network vulnerability assessment and safety evaluation application based on Bayesian theory. Int. J. Secur. Appl. **10**(12), 359–368 (2016). https://doi.org/10.14257/ijsia.2016.10.12.29

12. Wang, H., Chen, Z., Zhao, J., Di, X., Liu, D.: A vulnerability assessment method in industrial Internet of Things based on attack graph and maximum flow. IEEE Access **6**, 8599–8609 (2018). https://doi.org/10.1109/ACCESS.2018.2805690
13. George, G., Thampi, S.M.: A graph-based security framework for securing industrial IoT networks from vulnerability exploitation. IEEE Access **6**, 43586–43601 (2018) https://doi.org/10.1109/ACCESS.2018.2863244
14. He, W., Li, H., Li, J.: Unknown vulnerability risk assessment based on directed graph models: a survey. IEEE Access **7** 168201–168225 (2019). https://doi.org/10.1109/ACCESS2019.2954092
15. Wang, W., et al.: A vulnerability risk assessment method based on heterogeneous information network. IEEE Access **8**, 148315–148330 (2020)
16. Asvija, B., Eswari, R., Bijoy, M.B.: Bayesian attack graphs for platform virtualized infrastructures in clouds. J. Inf. Secur. Appl. **51**, 102455 (2020). https://doi.org/10.1016/j.jisa.2020.102455
17. Lallie, H.S., Debattista, K., Bal, J.: A review of attack graph and attack tree visual syntax in cyber security. Comput. Sci. Rev. **35,** 100219 (2020). https://doi.org/10.1016/j.cosrev.2019.100219
18. Vulnerability database. https://www.vuldb.com. Accessed 9 Feb 2022
19. CVSS v3.1 Specification. https://www.first.org/cvss/specification-document. Accessed 9 Feb 2022
20. CVSS PyPi. https://pypi.org/project/cvss/. Accessed 9 Feb 2022
21. CWE – Common Weakness Enumeration. https://cwe.mitre.org/. Accessed 9 Feb 2022
22. Download Metasploitable Intentionally Vulnerable Machine. https://information.rapid7.com/download-metasploitable-2017.html. Accessed 9 Feb 2022
23. Nessus Essential Vulnerability Scanner. https://www.tenable.com/products/nessus/nessus-essentials. Accessed 9 Feb 2022

Prediction of Thai Court Judgments in Criminal Cases Using Deep Neural Networks

Anawin Emarat, Bundit Manaskasemsak(✉), and Arnon Rungsawang

Massive Information and Knowledge Engineering Laboratory,
Department of Computer Engineering, Faculty of Engineering, Kasetsart University,
Bangkok 10900, Thailand
{anawin.e,bundit.m,arnon.r}@ku.th

Abstract. The Court of Justice currently applies information technology (IT) to support trials and to enhance operational efficiency. Various forms of innovations and information systems have been developed for operational responses, such as judgments, that require professionalism in trails and decisions of judges. Therefore, supporting information is important and necessary. This paper proposes a machine learning approach to support judges' decisions by predicting the outcome of a defendant's judgment in a criminal case. The profiles of previous defendants were thoroughly analyzed and presented with a set of features. Then, a deep neural networks (DNN) model was proposed to classify the judgment outcomes. We compared the performance of the proposed DNN with five traditional learning algorithms, i.e., decision tree, k-nearest neighbors, support vector machine, Naïve Bayes, and logistic regression. The results show that the proposed approach can predict judgment outcomes more accurately than the baseline approaches with an accuracy of 93.88%.

Keywords: Criminal law · Thai court · Judgment prediction · Deep learning · Neural networks · Machine learning

1 Introduction

Judging is not based only on the professionalism and experience of judges, but circumstances of the case are also a contributing factor. Currently, various technologies are being used in legal work. The Office of the Judiciary has brought technology to participate in collecting case data at every step from the prosecution process to the judgment of the case. This has become valuable information to support future judicial decisions. In predicting decisions, several factors, such as defendant's behavior and circumstances of the case, are needed to construct a learning model to classify judgments to support judges' decisions at trial.

A number of researchers have suggested concepts and processes of machine learning methods to classify legal information [1–5]. However, different countries tend to have different law enforcement. In addition, judgments are often based on

P. Meesad et al. (Eds.): IC2IT 2022, LNNS 453, pp. 111–118, 2022.
https://doi.org/10.1007/978-3-030-99948-3_11

the nature of the case, the defendant's behavior, and circumstances. These affect the definitions of related and different attributes for the classification. Therefore, the existing work may not be directly applied to Thai legal information as studied in our work.

This paper presents a deep learning approach to predict Thai Court judgments in criminal cases. To summarize, our contributions are as follows:

- We have thoroughly analyzed information on criminal cases obtained by the Office of the Judiciary. Then, we defined 48 characteristic features divided into 5 groups to represent each case.
- We proposed a deep neural networks (DNN) model to address the judgment prediction problem. Our results can help judges as a decision support system.
- We conducted the experiments on 73,394 criminal cases in Thailand, collected from December 1999 to October 2020. The results showed that our approach significantly outperformed traditional learning ones with accurate predictions.

The remainder of this paper is organized as follows. Section 2 reviews research studies related to ours. Section 3 details the defined features and the designed architecture of the proposed DNN. Section 4 reports performance evaluation comparing with five traditional learning methods. Finally, Sect. 5 concludes the paper.

2 Related Work

Many studies have used information technology to support the legal work in analyzing indications of offences. These can be divided into 2 types as follows.

2.1 Risk Assessment

Typically, personal risk assessment is an initial task that must be analyzed based on various indicators. Green [6] introduced a machine learning technique to analyze defendants' risk assessments using the COMPAS recidivism risk score algorithm. The study reported that factors affecting the defendant's score include behavioral background, environment, and case characteristics. Goel et al. [7] assessed the likelihood of armed suspects deterring their actions using logistic regression to predict crimes based on the level of risk with indicators such as gender, nationality, domicile, public transportation, accommodation, and nearby locations. Yoelao et al. [8] studied statistical assessment of the defendant's provisional release in cases related to sex, life, property, documents, traffic, and drugs. They found 14 risk indicators with statistically significant at 0.05. These include nationality, occupation, monthly income, domicile and current address, circumstances, denial of allegations, offence characteristics, impact on victims, opposition of provisional release, repeat offences, number of previous imprisonments, number of pended cases, number of accuses, and number of arrest warrants.

2.2 Trial

Court decisions are usually based on circumstances of the case. Berk and Bleich [9] used a random forest approach to predict the severity of cases from judicial decisions. The features used were divided into 4 categories: the defendant's personal background, such as gender and date of birth; juvenile case record data, such as the number of cases and the number of cases classified by gender; individual case record data, such as the number of cases and the number of crimes; and criminal history information, such as the number of days in prison, the number of probations, type of bails, and minimum age. Aletras et al. [2] used SVM to classify judicial decisions. Their study was divided into 3 subjects: against torture and other cruel, inhuman, or degrading treatment or punishment; rights to a fair trial; and respect Human Rights. Metsker et al. [5] studied the improvement of bills and the enforcement of judgments using data mining, cluster computing, and big data processing. The features include year of law issuance, penalty level, circumstances of the abatement, repeat offences, and confession.

Concerning judgments of Thai Court, Kowsrihawat et al. [4] proposed a deep learning model to predict the judgment in criminal cases of Thai Supreme Court. Sangkeettrakarn et al. [10] detected ambiguous words in Thai law text, determined by 3 conditions: a decision based on a judge's opinion, a decision based on evidence, and a decision based on other references. Krungklang and Sinthupinyo [11] studied diagnosis of words in Thai Criminal Code and brought for analysis of judicial decisions. The model was designed into 2 parts: rule generation and judgment classification with LSTM and BiLSTM.

3 Methodology

This section presents our effort to predict judicial decisions in criminal cases from Thai Court using a deep learning method. The components of the operational process are shown in Fig. 1. In the following, we will provide detailed information about the crime dataset, the feature extraction, and the architectural design of the model.

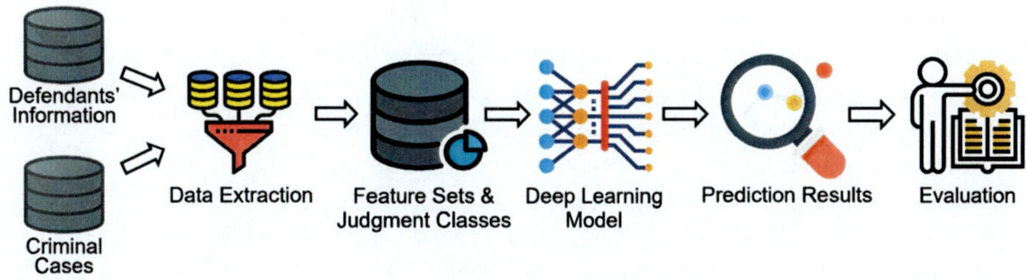

Fig. 1. The operational processes of the system.

3.1 Crime Dataset

We used a Thai case dataset received from the Office of the Judiciary. Within this dataset, only criminal cases whose judgments were complete between December 1999 and October 2020 were selected for analysis. So, we have a total of 73,394 cases. The crime data includes the defendants' personal information, behaviors, and case histories, as well as information about the trials and judgments.

3.2 Feature Extraction

Since our criminal records are virtually unstructured (i.e., stored in free-text format), we therefore used natural language processing (NLP) in conjunction with pattern recognition to extract distinctive features. After analyzing the crime data in detail, we have defined 48 features, divided into 5 categories as follows.

Category 1: *Defendant's background* refers to basic personal information and criminal history. This category consists of 12 features.

1. Gender, determined by male or female
2. Nationality, determined by Thai or foreigner
3. Occupation, determined by employed or unemployed
4. Age, determined by 26 years or below, 27–46 years, or 47 years and over
5. End level of the case, determined by investigation, trial, appeal, or petition
6. Number of detentions during the investigation
7. Number of detentions during the trial
8. Number of detentions during the appeal
9. Number of detentions during the petition
10. Imprisonment status during the trial, determined by yes or no
11. Bail status during the trial, determined by yes or no
12. Case ending status, determined by yes or no

Category 2: *Case details* refer to details of criminal cases related to current offences. This category consists of 9 features.

1. Crime type, determined by drugs, human trafficking, corruption, etc.
2. Number of charges against a crime
3. Number of particular charges related to larceny/fraud
4. Number of particular charges related to violence
5. Number of particular charges related to drugs
6. Number of particular charges related to weapons and immigration
7. Number of particular charges related to traffic and fuel
8. Amount of bail used to fight against the case
9. Victory, determined by prosecutor or defendant

Category 3: *Wrist* refers to sending a wrist associated with the cases classified by type and court hierarchy. This category consists of 9 features.

1. Number of appointments from the Court of First Instance
2. Number of summons from the Court of First Instance
3. Number of arrest warrants from the Court of First Instance
4. Number of appointments from the Court of Appeal
5. Number of summons from the Court of Appeal
6. Number of arrest warrants from the Court of Appeal
7. Number of appointments from the Supreme Court
8. Number of summons from the Supreme Court
9. Number of arrest warrants from the Supreme Court

Category 4: *Circumstance of the case* refers to characteristics of offences in each case by analyzing the defendant behavior, severity, case history, sued explanation, and brief judgment. This category consists of 9 features.

1. Number of pre-planned cases
2. Number of physical assault cases
3. Number of mental assault cases
4. Number of intimidation cases
5. Number of fraud cases
6. Number of negligence cases
7. Number of violence cases
8. Number of confession cases
9. Number of cases with sufferers

Category 5: *Case record* refers details of all related criminal cases, excluding the current case, by analyzing the defendant behavior, case history, sued explanation, and brief judgment. This category consists of 9 features.

1. Number of cases in which the defendant commits repeat offences
2. Number of co-defendants, excluding the defendant
3. Number of cases in which the defendant committed a criminal offense
4. Number of cases in which the defendant has been imprisoned before
5. Number of cases in which the defendant was opposed to provisional release
6. Number of cases in which the defendant has been charged
7. Number of cases that were not approved for provisional release
8. Number of cases bail pending trail
9. Number of cases in which the defendant escaped during the trial

After the extraction process, the feature set of some cases may not be complete. We therefore use the scikit-learn library[1] to proceed with the following steps. For features with numeric values, we replace the missing value with the mean value (i.e., Imputer), standardize the value by removing the mean and scaling to unit variance (i.e., StandardScaler), adjust the value to the unit norm from 0 to 1 (i.e., Normalizer), and scale using statistics to obtain unaffected value from outliers (i.e., RobustScaler). For features with categorical values, we

[1] https://scikit-learn.org/stable/.

replace the missing value with "none" by default. We also encode some features by splitting them into minor binary (0/1) columns based on the data values (i.e., one-hot encoding).

Finally, the class membership (i.e., judgment) is set to either imprisonment or release for each case.

3.3 Designed Deep Neural Networks Model

To deal with the issue of predicting judicial decisions, we propose a deep neural networks (DNN) model to classify each criminal case as either imprisonment or release based on its feature set. Figure 2 shows the conceptual model of the underlying DNN.

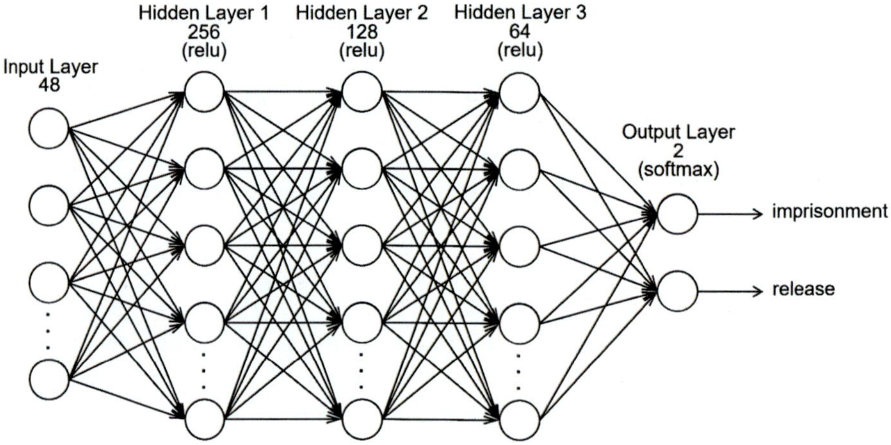

Fig. 2. The architecture of deep neural networks model.

Within the model, the input layer consists of the number of neurons based on the size of the input feature set (i.e., 48 features as described in Sect. 3.2). For the hidden layers, we have defined three layers with different number of neurons, which are 256, 128, and 64 neurons, respective. We also used the rectified linear unit (ReLU) as an activation function. To reduce the effects of overfitting, the dropout rate were set to 0.8, 0.4, and 0.2, respectively. Finally, we created the output layer with 2 neurons using the softmax activation function to predict a multinomial probability distribution so that the output value can be assigned to a class membership (i.e., imprisonment or release).

4 Performance Evaluation

We conducted the experiments on 73,394 criminal cases, of which 26,721 are prison cases and 46,673 are release cases. Based on this dataset, 70% of both prison and release cases were randomly selected for model training. The rest was used for testing.

We implemented our DNN model using the Python Keras with TensorFlow library[2, 3] according to the designed architecture and parameter settings described in Sect. 3.3. During the training phase, the number of epochs was fixed to 50 epochs. The Adam optimizer was used with a learning rate of 0.001 and a batch size of 128.

We compared our performance with that of five traditional learning algorithms: decision tree (DT), k-nearest neighbors (KNN), support vector machine (SVM), Naïve Bayes (NB), and logistic regression (LR), respectively. Table 1 shows the compared results reported in terms of four standard metrics: precision, recall, F-measure, and accuracy.

Table 1. Comparison of prediction performance between different learning algorithms.

Models	Imprisonment			Release			Accuracy
	Precision	Recall	F-Measure	Precision	Recall	F-Measure	
DNN	**93.28**	**90.23**	**91.73**	**94.22**	**96.08**	**95.14**	**93.88**
DT	72.28	81.52	76.63	90.62	85.10	87.77	83.94
KNN	73.02	78.35	75.59	88.45	85.13	86.76	82.83
SVM	67.77	73.58	70.56	86.07	82.35	84.17	79.41
NB	77.56	66.93	71.86	78.06	85.87	81.78	77.88
LR	62.22	56.65	59.30	72.74	77.08	74.85	68.91

All values are reported as percentages. Bold text emphasizes the maximum value

As can be seen, our DNN model provides scores significantly higher than the others in all cases. These reported values are also very high, indicating that DNN can accurately predict almost all cases of imprisonment and release. In other words, the 93.88% accuracy of DNN summarizes the best prediction performance.

5 Conclusion

This work presents a deep learning approach for predicting judicial decisions being applicable to criminal cases in Thailand. Forty-eight effective features have been defined to represent each crime record sentenced imprisonment or release. Experiments were conducted on a crime dataset received from the Office of the Judiciary. The results proved that our efforts outperform all traditional learning algorithms with highly accurate prediction scores.

There are still many factors such as other features, optimizers, and learning rates, to be investigated that we expect to improve the performance of our model

[2] https://keras.io.

[3] https://www.tensorflow.org.

in future work. An automatic fine-tuning of parameters such as the number of neurons is also an important task for practical use with other datasets.

Acknowledgement. We thank to the Office of the Judiciary for providing the crime dataset used in this work.

References

1. Thammaboosadee, S., Watanapa, B., Charoenkitkarn, N.: A framework of multi-stage classifier for identifying criminal law sentences. Procedia Comput. Sci. **13**, 53–59 (2012)
2. Aletras, N., Tsarapatsanis, D., Preoţiuc-Pietro, D., Lampos, V.: Predicting judicial decisions of the European court of human rights: a natural language processing perspective. PeerJ Comput. Sci. **2**, e93 (2016)
3. Katz, D.M., Bommarito II, M.J., Blackman, J.: A general approach for predicting the behavior of the supreme court of the united states. PLoS ONE **12**(4), e0174698 (2017)
4. Kowsrihawat, K., Vateekul, P., Boonkwan, P.: Predicting judicial decisions of criminal cases from thai supreme court using bi-directional GRU with attention mechanism. In: Proceedings of the 5th Asian Conference on Defense Technology, pp. 50–55 (2018)
5. Metsker, O., Trofimov, E.V., Petrov, M., Butakov, N.: Russian court decisions data analysis using distributed computing and machine learning to improve lawmaking and law enforcement. Procedia Comput. Sci. **156**, 264–273 (2019)
6. Green, B.: "Fair" risk assessments: a precarious approach for criminal justice reform. In: Proceedings of the 5th Workshop on Fairness, Accountability, and Transparency in Machine Learning (2018)
7. Goel, S., Rao, J.M., Shroff, R.: Personalized risk assessments in the criminal justice system. Am. Econ. Rev. **106**(5), 119–123 (2016)
8. Yoelao, D., Peungphosop, N., Junprasert, T., Chaiakkarakan, N., Pisitchinda, T.: Enhancing Effectiveness in Decision among Pretrial Release Defendants: A Behavioral Science Approach to Risk Management. Srinakharinwirot University (2018)
9. Berk, R., Bleich, J.: Forecasts of violence to inform sentencing decisions. J. Quant. Criminol. **30**, 79–96 (2014)
10. Sangkeettrakarn, C., Haruechaiyasak, C., Theeramunkong, T.: Fuzziness detection in thai law texts using deep learning. In: Proceedings of the 10th International Conference of Information and Communication Technology for Embedded Systems (2019)
11. Krungklang, W., Sinthupinyo, S.: An analysis of natural language text relating to thai criminal law. In: Proceedings of the 12th International Conference on Electronics, Computers and Artificial Intelligence (2020)

Multi-objective Optimization of Aluminum Profile Extrusion Process

Jaramporn Hassamontr[1] (✉) and Theera Leephaicharoen[2]

[1] King Mongkut's University of Technology North Bangkok, Bangkok 11000, Thailand
jaramporn@gmail.com

[2] Thai Metal Aluminum Company Limited, Muang 10280, Samutprakarn, Thailand
theera@thaimetal.com

Abstract. The purpose of this research is to develop an optimization tool to support planning of aluminum profile extrusion process. It is based on two generalized extrusion setups namely multiple extrusions per billet and multiple billets per extrusion. The objectives of the planning are to ensure that both resulting yield and productivity are close to maximum levels. An enumeration technique is used to identify all feasible solutions of which optimal solutions with respect to each objective highlighted. This allows human process planners to select a solution to meet customer order on a case-by-case basis and to evaluate long-term hardware improvement potentials. The effectiveness of this tool is demonstrated using case studies from an industry.

Keywords: Multi-objective optimization · Aluminum profile extrusion · Yield optimization · Throughput maximization · Decision support system

1 Introduction

While companies strive to improve their operations, applying optimization techniques in industrial settings can be difficult due to many reasons. For one, in most industrial settings there is more than one objective. For others, it may be difficult to formulate constraints mathematically. Furthermore, some constraints are changed from one operation to the next. Thus, most companies rely on human experts to perform planning or use some types of decision support system to assist human planners. In aluminum profile extrusion industry, such a decision support system is proposed and shown in Fig. 1. Each new product goes through the stage of die design, validation and testing on an extrusion press machine. This is usually done by research and development personnel to determine proper extrusion process parameters. These process parameter data are kept in die and extrusion press database. Dashed line represents boundary for the decision support system. When the process planner receives an order by a customer, he needs to select the extrusion press and generate an operation plan for the extrusion press operator. When the operator executes the plan, he may need to adjust the operation and process parameters so as to meet desired quality and quantity while keeping the cost low. The new process parameters are recorded in production database.

P. Meesad et al. (Eds.): IC2IT 2022, LNNS 453, pp. 119–131, 2022.
https://doi.org/10.1007/978-3-030-99948-3_12

This research is an extension from past research by Hassamontr and Leephaicharoen [1] where an objective of maximizing yield was considered in planning aluminum profile extrusion process. The manufacturing operation was categorized into two distinct setups-multiple extrusions per billet and multiple billets per extrusion. In this research, another objective of maximizing throughput is added. Based on time study, productivity models for each extrusion setup are developed. While it is possible to employ a traditional approach of multi-objective optimization to solve the problem, this research focuses on the development of preliminary decision support system where human planners are presented with all feasible solutions with optimal solution(s) highlighted. This allows them to (1) select the most suitable operation themselves for each customer order, (2) provide the backup plan for front-line operators to adjust the operation should some process parameters are accidentally changed. The optimization tool developed can be used to identify the impact of long-term hardware improvements on operation performances also. These benefits are illustrated using case studies.

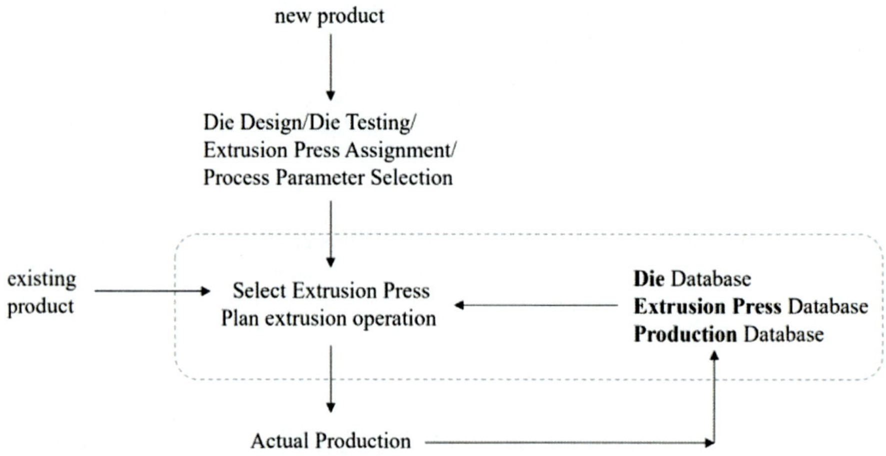

Fig. 1. Schematics for decision support system for aluminum profile extrusion process.

2 Literature Review

Aluminum profile extrusion is a very established industry. Its global trade value is more than 57 billion US dollars in 2020 [2]. Aluminum Extruders Council [3] is a very active trade organization promoting the application of aluminum alloys. Physical modeling of aluminum extrusion is summarized by Sheppard [4]. The hardware part has been continuously improved by machine manufacturers. In terms of process optimization, most research focused on minimizing scrap. Tabucanon and Treewannakul [5] first introduced the formulation of extrusion scrap minimization for pre-cut billet sizes. Masri and Warburton [6, 7] considered global optimization of pre-cut billet sizes applied to all past product orders. Various sources of scrap were considered by Hajeeh [8]. Oza and Gotowala [9] conducted experimental studies to investigate the effect of process parameters on scrap.

Manufacturing companies tend to develop the optimization tool in-house. These tools are not publicized nor accessible to smaller companies. This research presents a preliminary attempt to develop a decision support system for aluminum profile extrusion planning based on previous work by Hassamontr and Leephaicharoen [1]. Research on decision support system is well surveyed by Eom and Kim [10]. Duttta, Gupta and Fourer [11] reported the application of optimization-based decision support system application in aluminum sheet manufacturing industry at global level while Vidal, Alberti, Ciurana and Casedesus [12] applied the decision support system on process parameter selection in machining industry.

Another aspect where this research will be different from past research is that the objective of the optimization is expanded to include productivity, thus making the problem multi-objective. The methodology for multi-objective optimization has been well-studied by several researchers, such as Giagkiosis and Fleming [13], Marler and Arora [14]. However, with the decision support approach in mind, feasible solution space will be presented to human process planners instead. To assist the selection of process parameters, optimal solutions for each objective and pareto-optimal solutions are highlighted.

3 Research Methodology

3.1 Problem Formulation

Aluminum profile extrusion is a manufacturing process to make long extruded uniform cross section bars. These bars are mostly used as structural supports for machineries, building structure, car frame and electronic components, such as heat sink in computers. The most widely used material for the process is aluminum alloy. General manufacturing operation for extruded profile is shown in Fig. 2. An aluminum log, usually of standard size, is cut into shorter billets. Each billet is heated, placed into a chamber and pushed through a die. Extruded profiles are pulled out from the other side of the die until they reach desired length. Then the pulling stops so that the saw can cut profiles to this desired length. The profiles are then pulled apart by grippers to straighten and cut to shorter lengths that customer requires. There are a lot of process parameters involved in the operation. Mostly they are determined in the prototyping stage. The process parameters are specific to each product cross section, the die used and even the extrusion press. Changing any one of these components requires process parameter verification and adjustments.

The following are process parameters pre-determined in the prototyping stage. They are known in advance before actual mass production.

- Aluminum log: weight G, in kg and its density b, in kg/cm.
- Extrusion press constraints: maximum and minimum billet sizes that can be put into chamber B_{max} and B_{min} respectively, minimum amount of backend remained after extrusion, e, all in kg.
- Extrusion die: number of openings in the die, a, amount of material remained in die pot P, in kg, and profile density w, in kg/m

- Runout table: cutting saw position L_s from the die, maximum and minimum extrusion lengths that can be supported by runout table R_{max} and R_{min} respectively, gripper sizes h and t to straighten profiles.

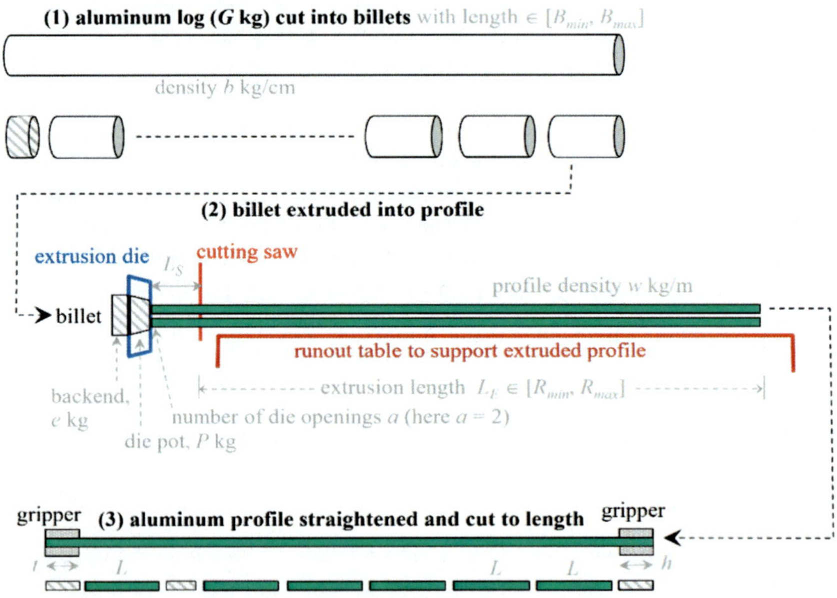

Fig. 2. Aluminum profile extrusion process.

The parameters need to be revised for each operation are customer order length L and required quantity n. As one can see in Fig. 2, there are many sources of material scraps generated in every step from converting aluminum log into customer-specified lengths. And, as it turns out, it is not the only thing production planner is concerned about. Companies looks to improve their throughput as well.

Broadly speaking there are two ways to operate an extrusion press. If a relatively large billet size is extruded, each of its resulting extrusion lengths, as cut by the saw, must be within R_{min} and R_{max}. This leads to multiple extrusions per billet, with single extrusion as a special case. On the other hand, when a billet size is relatively small, more than one billet is needed to create an extrusion length between R_{min} and R_{max}. This is called multiple billets per extrusion. Both configurations have different scrap rate and require different cycle time.

Before diving into mathematical specifics, let us summarize assumptions made in this study.

- Process parameters are known or pre-determined.
- Saw used for profile cutting on the runout table is at fixed position L_s from the die.
- Material losses due to cutting saw is negligible.

- Average aluminum profile density, w, is pre-estimated. Usually, a value close to the upper range is used to ensure that enough material remained in the chamber after each extrusion.
- In multiple extrusions per billet setup, for the sake of simplicity in actual operation, extrusion length is assumed constant. At most two billet sizes are used to meet a customer order. This will be discussed in Sect. 3.2.
- For multiple billets per extrusion setup, extrusion length should be uniform for each set of billets. At most three billet sizes are used to form a billet set. This will be discussed in Sect. 3.3.

3.2 Multiple Extrusions Per Billet

The extrusion operation usually starts with an operator installing new die into the extrusion press. Technical parameters relevant to the die are retrieved and input to the machine. The first billet is loaded into the chamber. Let us consider this as stage (1), starting point of the cycle. Time elements for two extrusions per billet are demonstrated in Fig. 3.

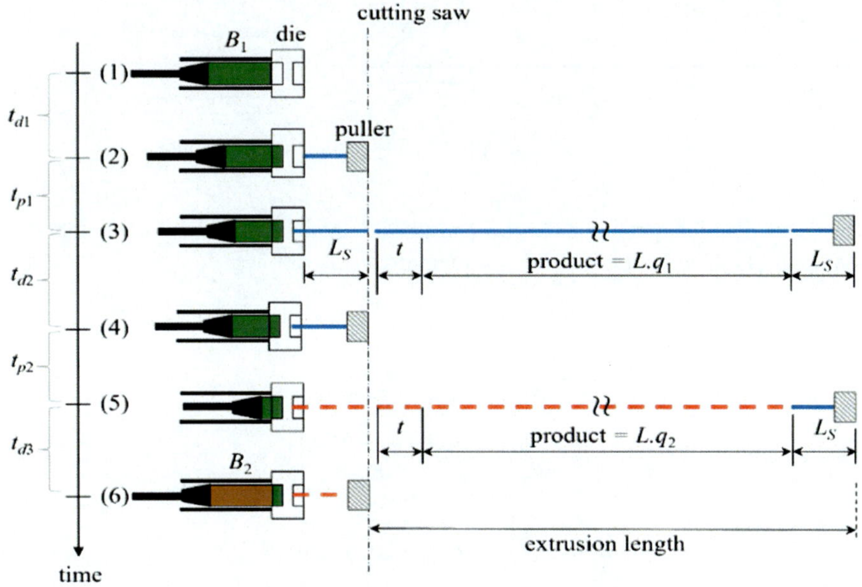

Fig. 3. Time elements for two extrusions per billet ($C_1 = 2$).

Delay time between when the billet is ready in the chamber and when aluminum profile starts to come out of the die is denoted by t_{d1} in stage (2). The extrusion proceeds until the puller reaches specified extrusion length, as marked by stage (3). This time duration is denoted by t_{p1}. Cutting saw cut profiles while the puller moves to the cutting saw as shown in stage (4). This delay time is t_{d2}.

Once again, the puller starts to drag profiles until it reaches the specified extrusion length in stage (5), this takes t_{p2}. At this point the remaining billet in the chamber can no longer be extruded. Several things happen at the same time. As before, the saw cuts profiles while the puller moves back to the front of the die. The remaining back end of the billet is removed from the chamber and the next billet is loaded into the chamber. This is stage (6) in Fig. 3. All these overlapping events take t_{d3}. Then the cycle repeats until order quantity is reached.

The extrusion speed is usually constant with ramp-up and ramp-down times in the beginning and the end of each extrusion. The extrusion speed profile along with its normal data is shown in Fig. 4.

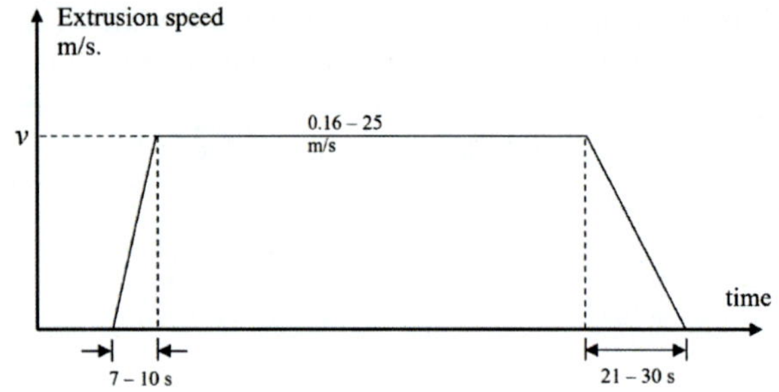

	ext/blt	blt/ext
td1 (s)	5	5
td2 (s)	19	19
td3 (s)	26	26
td4 (s)		10
ramp-up time (s)	10	10
ramp-down time (s)	30	30
total ramp distance (m)	4	4
ext speed (m/s)	0.2	0.2

Fig. 4. Extrusion speed profile and specific time constants.

The optimization problem can be formulated in a similar fashion to that from Hassamontr [1]. Constraints for the optimization are as follows.

$$B_1 = e + P + aw\{C_1(L_s + Lq_1 + t) + L_s + h\} \tag{1}$$

$$B_2 = e + aw\{C_2(t + L_s + Lq_2) + h\} \tag{2}$$

$$B_{min} \leq B_{1,2} \leq B_{max} \tag{3}$$

$$L_{E1,2} = L_s + t + Lq_{1,2} + h \tag{4}$$

$$R_{min} \leq L_{E1,2} \leq R_{max} \tag{5}$$

$$n_{b2} = \left\lceil \frac{n - aC_1q_1}{aC_2q_2} \right\rceil \tag{6}$$

$$g_1 = \left\lfloor \frac{G - B_1/b}{B_2/b} \right\rfloor \tag{7}$$

$$g_2 = \left\lfloor \frac{G}{B_2/b} \right\rfloor \tag{8}$$

$$M_2 = \left\lceil \frac{n_{b2} - g_1}{g_2} \right\rceil \tag{9}$$

$$t_{p2} = 10 + \frac{L_{E2} + L_S - 4}{0.2} + 30 \tag{10}$$

$$C_1, C_2 \geq 1, integer \tag{11}$$

$$q_1, q_2, n_{b2}, g_1, g_2, M_2, \geq 0, integer \tag{12}$$

$$B_1, B_2, L_{E1}, L_{E2}, t_{p2} \geq 0 \tag{13}$$

Constraints (1)–(3) are mass balance equations for the first and continuing billets which can be derived from Fig. 3. Each billet must be within B_{min} and B_{max}. Equations (4)–(5) are to ensure that the first and subsequent extrusion lengths are within runout table limits. The number of consecutive billets required, n_{b2}, to meet customer order quantity can be calculated from (6). The total number of consecutive billets, g_1 and g_2, that can be packed into the first and next aluminum logs are calculated in (7) and (8). Equation (9) is used to determine how many aluminum logs, M_2, aside from the first log, are used to meet customer order quantity. The extrusion time t_{p2} can be calculated in (10) based on data in Fig. 4. All decision variables are denoted in (11)–(13).

It should be noted from the assumption that only the first extrusion of the very first billet may have a unique extrusion length L_{E1}. All subsequent extrusions will have identical extrusion length L_{E2}. Thus the cycle time of each billet can be calculated, based on t_{p2}, by $t_{d1} + C_2 t_{p2} + (C_2 - 1)t_{d2} + t_{d3}$. Each billet generates aC_2q_2 pieces per cycle. Two objectives for this optimization problem can be expressed as maximizing the yield as shown in (14) and maximizing throughput in pieces per hour as shown in (15). Here yield is the ratio of total product weight made when aluminum logs are used up to exceed customer demand quantity.

$$maximize \qquad 100 * Lw \frac{aC_1q_1 + aC_2q_2(g_1 + g_2M_2)}{Gb(1 + M_2)} \tag{14}$$

$$maximize \qquad 3600 * (aC_2q_2)/\left\{t_{d1} + C_2t_{p2} + (C_2 - 1)t_{d2} + t_{d3}\right\} \tag{15}$$

3.3 Multiple Billets Per Extrusion

Operating multiple billets per extrusion is very similar to that of multiple extrusions per billet, except that more than one billet is loaded in a cycle. Figure 5 demonstrates time elements for three billets per extrusion cycle. To simplify the notations, the subscript remained the same as in Sect. 3.2, but the prime is added to denote multiple billets per extrusion case. Loading time for the next billet, other than the first billet is t'_{d4}. Typical delay time as well as ramp-up and ramp-down times for multiple billets per extrusion are shown in Fig. 4.

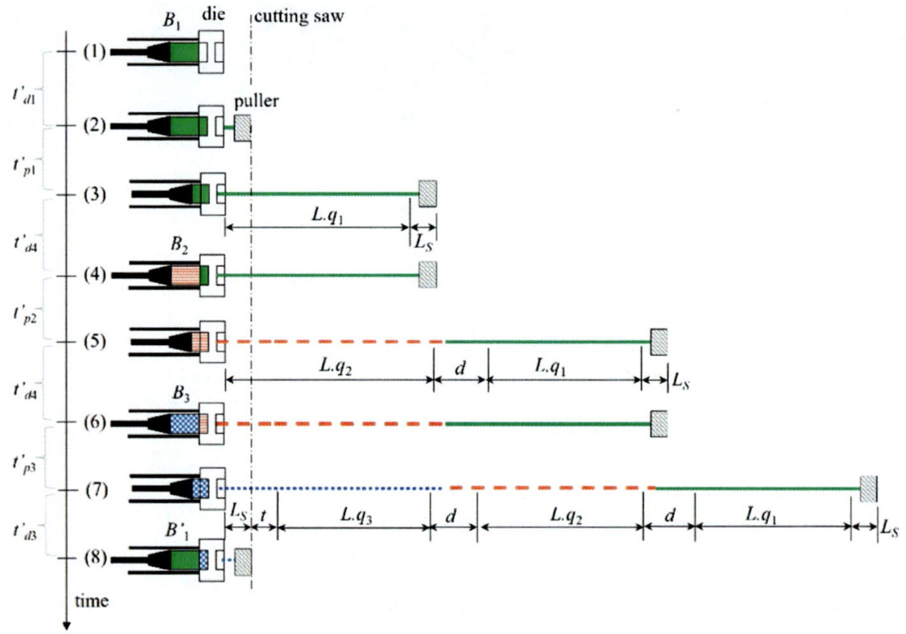

Fig. 5. Time elements for three billets per extrusion ($D_2 = 1$).

The optimization problem can be formulated with all constraints as follows.

$$B_1 = e + P + aw\{Lq_1 + L_s + h\} \tag{16}$$

$$B_2 = e + P + aw\left\{Lq_2 + \left(d - \frac{P}{aw}\right)\right\} \tag{17}$$

$$B_3 = e + P + aw\left\{L_s + t + Lq_3 + \left(d - \frac{P}{aw}\right)\right\} \tag{18}$$

$$B_1^{'} = e + P + aw\left\{Lq_1 + \left(d - \frac{P}{aw}\right)\right\} \tag{19}$$

$$B_{min} \leq B_{1,2,3}, B_1^{'} \leq B_{max} \tag{20}$$

$$L_E = (L_s + Lq_1) + D_2(d + Lq_2) + (d + Lq_3 + t) \tag{21}$$

$$R_{min} \leq L_E \leq R_{max} \tag{22}$$

$$n_s = \left\lceil \frac{n - a(q_1 + D_2q_2 + q_3)}{a(q_1 + D_2q_2 + q_3)} \right\rceil \tag{23}$$

$$h_1 = \left\lfloor \frac{G - (B_1 + D_2B_2 + B_3)/b}{\left(B_1^{'} + D_2B_2 + B_3\right)/b} \right\rfloor \tag{24}$$

$$h_2 = \left\lfloor \frac{G}{(B_1' + D_2 B_2 + B_3)/b} \right\rfloor \tag{25}$$

$$N_2 = \left\lceil \frac{n_s - h_1}{h_2} \right\rceil \tag{26}$$

$$T = t_{d1}' + \left(t_{p1} + t_{d4}' + t_{d1}'\right) + D_2\left(t_{p2} + t_{d4}' + t_{d1}'\right) + \left(t_{p3} + t_{d3}'\right) \tag{27}$$

$$B_1, B_1', B_2, B_3, T \geq 0 \tag{28}$$

$$D_2, q_1, q_2, q_3, n_s, h_1, h_2, N_2 \geq 0, \textit{integer} \tag{29}$$

Equations (16)–(20) are mass balance equations for each billet which can be derived from Fig. 5 where each billet is bounded by B_{min} and B_{max}. Equations (21)–(22) are to impose constraint on extrusion length for each billet set. The number of subsequent billets required, n_s, to meet customer order quantity can be calculated from (23). The total number of subsequent billet sets that can be packed into the first and next aluminum logs, h_1 and h_2, are calculated in (24)–(25). Equation (26) is used to determine how many aluminum logs, aside from the first, are used to meet or exceed customer order quantity. The cycle time for each billet set, T, is calculated in (27), where extrusion time $t_{p1,2,3}'$ is calculated in a similar fashion as Eq. (10). All decision variables are denoted in (28)–(29).

Similarly, two objective functions (30) and (31) are to maximize the yield and production rate respectively.

$$\textit{maximize} \qquad 100 * Lw \frac{a(q_1 + D_2 q_2 + q_3)(1 + h_1 + N_2 h_2)}{Gb(1 + N_2)} \tag{30}$$

$$\textit{maximize} \qquad 3600 * a(q_1 + D_2 q_2 + q_3)/T \tag{31}$$

3.4 Solution Procedure

While this can be formulated as a multi-objective optimization problem, it is difficult to solve using commercial software packages as it is non-linear and discrete. An enumeration technique is devised to solve this multi-objective optimization problem. Proposed algorithm can be described as follows.

- First, enumerate independent variables to identify all feasible billet sizes $\in [B_{min}, B_{max}]$, extrusion lengths $\in [R_{min}, R_{max}]$ and resulting objective values. All feasible solutions are tabulated for easy search.
- Highlight optimal solution for each objective values as well as pareto-optimal solutions, presented visually on 2-dimensional feasible solution space.
- Let human planner decide on the operation plan, instead of imposing on tradeoffs between the two objectives.

The optimization tool was first prototyped using Visual Basic Application in Microsoft Excel.

4 Case Studies

In order to illustrate the optimization tool's capability, an example data was taken from industry. The input data for the program is shown in Table 1. Unless specified otherwise, all case studies mentioned in this section will be based on these process parameters.

Table 1. Parameters used in case studies.

Aluminum log	$G = 387.44$ kg (length 5.8 m), $b = 0.668$ kg/cm
Customer order	$L = 3.1$ m, $n = 1800$ pcs
Extrusion press setup	$B_{min} = 16.7$ kg (25 cm), $B_{max} = 53.44$ kg (80 cm), $e = 1.75$ kg
Die	$a = 2$, $w = 0.545$ kg/m, $P = 1.25$ kg
Runout table	$R_{min} = 20$ m, $R_{max} = 50$ m, $L_s = 1.5$ m, $h = 1$ m, $t = 1$ m, $d = 1$ m

4.1 Feasible Solution Space

To make 1800 pieces of 3.1 m-long profile, the program generates feasible solutions for both multiple extrusions per billet and multiple billets per extrusion setups. All feasible solutions are presented in a tabulated form with the optimal solution for each objective highlighted. To simplify the visual display of feasible solutions, the solutions are categorized based on extrusion setups and the number of pieces made per cycle, which depends on whether the cycle is represented by a billet or by an extrusion. Figure 6 illustrates these feasible solutions based on pieces per cycle and their objective values.

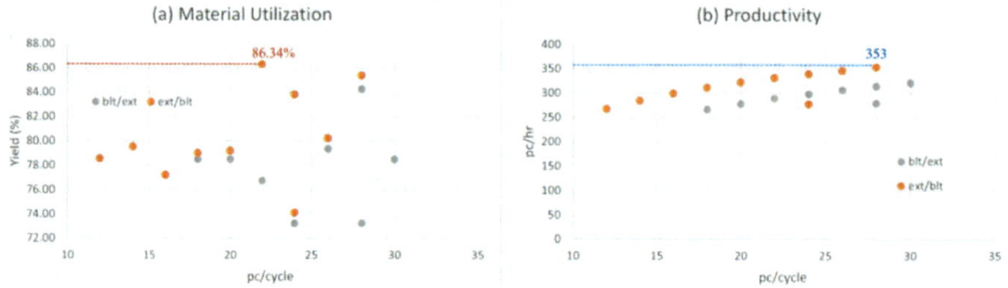

Fig. 6. Feasible solutions for each objective.

Any feasible solutions with the same pieces per cycle, same extrusion setup and same objective values were removed for clarity. Furthermore, those inferior solutions were removed. Based on Fig. 6, multiple extrusions per billet setup provides better performance in terms of yield and productivity, even though multiple billets per extrusion setup may be able to provide higher number of pieces per cycle. The maximum yield achieved is 86.34% where a single extrusion per billet is used with $q_1 = q_2 = 11$ pieces of product per profile. The solution results in 331 pieces per hour. On the other hand, if one considers productivity objective, the maximum production rate is 353 pieces per hour. There are such four solutions but the one with highest yield is 85.38% where single extrusion per billet is performed with $q_1 = 13$ and $q_2 = 14$ pieces of product per profile. Both solutions are quite comparable, and user can select either one based on his subjective criteria.

For illustration purposes, suppose the process planner decides to go for the highest productivity approach since the yield is relatively close to the highest yield achieved already. When the plan is released to shop floor, some process parameters used in the calculation may be changed. For example, billet weight may be less than what is planned, 0.668 kg/cm, or profile density may be higher than planned, 0.545 kg/m. Such variations will lead to different extrusion length L_E or remaining backend e. The machine operator can use a table as shown in Fig. 7 below to adjust for the new extrusion length as well as cutting profile without having to re-run the optimization tool.

q1	c1	B1 (cm)	q2	c2	B2 (cm)	ext_length2 (m)	ext_length 1 (m)	total pcs	total billet wght (kg)	total product wght(kg)	Yield_ log(%)	pc/hr C1
12	1	72.98	12	1	68.66	40.70	40.70	1920	3874.40	3243.84	83.72	339
12	1	72.98	13	1	73.72	43.80	40.70	1818	3874.40	3071.51	79.28	346
12	1	72.98	14	1	78.77	46.90	40.70	1956	3874.40	3304.66	85.29	353
13	1	78.03	12	1	68.66	40.70	43.80	1922	3874.40	3247.22	83.81	339
13	1	78.03	13	1	73.72	43.80	43.80	1820	3874.40	3074.89	79.36	346
13	1	78.03	14	1	78.77	46.90	43.80	1958	3874.40	3308.04	85.38	353

Fig. 7. Tabulated results for machine operator to adjust the operation.

4.2 Effect of Improving Maximum Billet Size

Feasible solutions shown in Fig. 6 can be presented in a feasible solution space where two resulting objectives are plotted with each other as shown in Fig. 8(a). From the figure, it is easy to identify the optimal solution for each objective. The red line represents the yield where we want to move as highest as possible while the blue line represents the productivity where we want to move to the right as much as possible. With this approach, one can easily identify optimal solution for each objective and allow users to make tradeoff decision based on their subjective preferences.

The optimization tool developed can be used to quantify potential long-term hardware improvement. For example, one might ask what would happen to productivity and yield if billet chamber is made larger than before? Fig. 8(b) demonstrates feasible solution space when the upper bound of billet size B_{max} is increased from 80 cm to 120 cm. Only the solutions from multiple extrusions per billet are presented in the graph. From the figure, the highest yield increases to 87.21% where the productivity remains the same. The maximum productivity increases from 353 pc/h to 358 per hour with the yield of 79.54%. Here there exists a third alternative, pareto-optimal solution where both objectives are compromised with 353 pc/h and 85.64% yield.

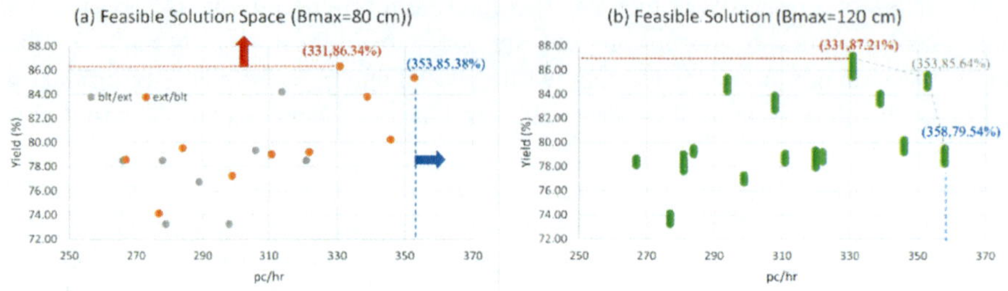

Fig. 8. Feasible solution space for (a) $B_{max} = 80$ cm and (b) $B_{max} = 120$ cm.

4.3 Effect of Improving Maximum Runout Table Length

In this case study, another possible hardware improvement is considered. By extending the runout table length, ie. increasing R_{max}, more products can be made in each extrusion. In order to see the impact of increasing R_{max} more clearly, B_{max} is set at 120 cm. Once again, feasible solution space for different R_{max} can be investigated. Feasible solution for $R_{max} = 50$ m is shown in Fig. 8(b) whereas those for $R_{max} = 63$ and 75 m are presented in Fig. 9(a) and 9(b) respectively. By increasing R_{max}, the maximum yield increases from 87.21% to 89.15% and 91.57%. Productivity increases from 358 to 376 and 386 pc/hr. Improving both B_{max} and R_{max} at the same time can drastically improve operating performances. It should be noted that, increasing R_{max} from 50 m to 63 and 75 m while holding B_{max} at 50 cm will not expand feasible solution space.

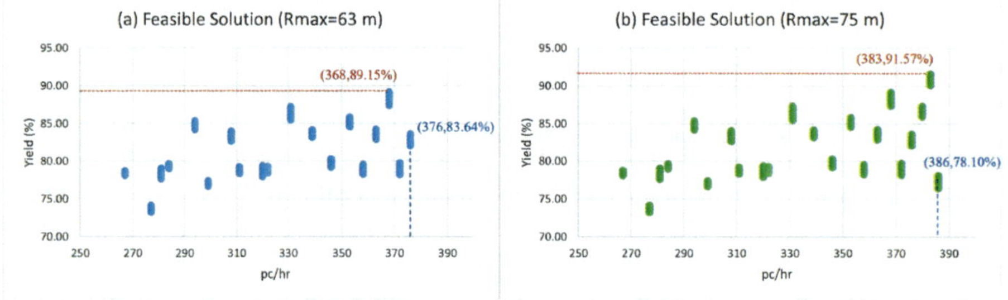

Fig. 9. Feasible Solution Space when $B_{max} = 120$ cm for (a) $R_{max} = 63$ m and (b) $R_{max} = 75$ m.

5 Conclusion

In this research, an optimization tool is developed to assist human planners in designing aluminum profile extrusion operations. The system considers two general ways to operate an extrusion press, namely multiple extrusions per billet and multiple billets per extrusions. The objectives are to maximize both material yield and productivity simultaneously.

Its visual display of feasible solution space allows process planners to make decision more objectively. Its tabulated solution allows machine operators to adjust operation upfront without the need to re-optimize if process parameters are inadvertently changed. Furthermore, it can be used to identify the impact of long-term hardware improvements, such as maximum billet size and maximum runout table length, on performance measures. These capabilities were demonstrated using various case studies. More work can be done to extend this optimization tool into a decision support system where multiple extrusion presses are considered simultaneously.

References

1. Hassamontr, J., Leephaicharoen, T.: Modeling of aluminum profile extrusion yield: online billet cutting. In: Meesad, P., Sodsee, S., Jitsakul, W., Tangwannawit, S. (eds.) IC2IT 2021. LNNS, vol. 251, pp. 233–243. Springer, Cham (2021). https://doi.org/10.1007/978-3-030-79757-7_23
2. Grand View Research. https://www.grandviewresearch.com/industry-analysis/aluminum-extrusion-market. Accessed 09 Feb 2022
3. Aluminum Extruder Council. https://www.aec.org/default.aspx. Accessed 09 Feb 2022
4. Sheppard, T.: Extrusion of Aluminum Alloys. Springer, Dordrecht (1999). https://doi.org/10.1007/978-1-4757-3001-2
5. Tabucanon, M.T., Treewannakul, T.: Scrap reduction in the extrusion process: the case of an aluminium production system. Appl. Math. Model. **11**, 141–145 (1987). https://doi.org/10.1016/0307-904X(87)90158-2
6. Masri, K., Warburton, A.: Optimizing the yield of an extrusion process in the aluminum Industry. In: Tamiz, M. (ed.) Multi-Objective Programming and Goal Programming. LNE, vol. 432, pp. 107–115. Springer, Heidelberg (1996). https://doi.org/10.1007/978-3-642-87561-8_9
7. Masri, K., Warburton, A.: Using optimization to improve the yield of an aluminum extrusion plant. J. Oper. Res. Soc. **49**(11), 1111–1116 (1998). https://doi.org/10.1057/palgrave.jors.2600616
8. Hajeeh, M.A.: Optimizing an aluminum extrusion process. J. Math. Stat. **9**(2), 77–83 (2013). https://doi.org/10.3844/jmssp.2013.77.83
9. Oza, V.G., Gotowala, B.: Analysis and optimization of extrusion process using Hyperworks. Intl. J. Sci. Res. Dev. **2**(8), 441–444 (2014)
10. Eom, S., Kim, E.: A survey of decision support system applications (1995–2001). J. Oper. Res. Soc. **57**, 1264–1278 (2006). https://doi.org/10.1057/palgrave.jors.2602140
11. Dutta, G., Gupta, N., Fourer, N.: An optimization-based decision support system for strategic planning in a process industry: the case of aluminium company in India. J. Oper. Res. Soc. **62**, 616–626 (2011). https://doi.org/10.1057/jors.2010.8
12. Vidal, A., Alberti, M., Ciurana, J., Casadesus, M.: A decision support system for optimising the selection of parameters when planning milling operations. Intl. J. Mach. Tool Manuf. **45**, 201–210 (2005). https://doi.org/10.1016/j.ijmachtools.2004.07.007
13. Giagkiozis, I., Fleming, P.J.: Methods for multi-objective optimization: an analysis. Inf. Sci. **293**, 1–16 (2015). https://doi.org/10.1016/j.ins.2014.08.071
14. Marler, R.T., Arora, J.S.: The weighted sum method for multi-objective optimization: new insights. Struct. Multidisc. Optim. **41**, 853–862 (2010). https://doi.org/10.1007/s00158-009-0460-7

Bitcoin Considered Harmful

Wolfgang A. Halang(✉) and Mario M. Kubek

FernUniversität in Hagen, 58084 Hagen, Germany
{wolfgang.halang,mario.kubek}@fernuni-hagen.de

Abstract. The crypto currency Bitcoin is rooted in peer-to-peer computing. To critically assess the feasibility of this approach, first the concept of Bitcoin is outlined. Then, widely based on original citations of its defining document, a discussion of its properties leads to the result that this crypto currency is inherently insecure and not amenable to real-time operation, as well as that the generation of Bitcoins and the processing of payment transactions wastes resources to a not justifiable extent. Hence, Bitcoin is extremely inadequate as an electronc payment system, in particular for worldwide use on a daily basis.

Keywords: Bitcoin · Crypto currency · Peer-to-peer computing · Distributed systems · Cryptography

1 Outline of the Crypto Currency Bitcoin

An author unknown to the present day published in 2008 under the pseudonym Satoshi Nakamoto the white paper [1], in which he presented his concept of the electronc payment system Bitcoin, that is by now also eponymous for the world's most well-known crypto currency (abbreviation: BTC).

The Bitcoin system was developed to process—without intermediaries — worldwide payments via insecure communication channels such as the Internet. Thus, the customary model of trust was to be replaced, which is based on financial institutions as money transfer operators and trustworthy third parties, which entails transaction costs by commission fees and costs for protection mechanisms, and which enables to reverse transactions. Instead, financial transactions were to be made irreversible in the Bitcoin system. To this end, they are to be confirmed in their chronological order by computational results called Proof-of-Work prepared by computer nodes participating in a Peer-to-Peer (P2P) network, i.e. in a purely distributed environment. At the same time, the problem of multiple spending of Bitcoins—improperly called double-spending by Nakamoto—is to be solved by this approach, only under the condition, however, that more than half of the nodes acts "honestly".

The crypto currency Bitcoin has no inherent recoverability due to its lack of any backing in real assets deposited somewhere or in natural resources. The computational Proofs-of-Work are, apart from the market value resulting from supply and demand, and the maximum amount of BTCs limited by the Bitcoin

P. Meesad et al. (Eds.): IC2IT 2022, LNNS 453, pp. 132–141, 2022.
https://doi.org/10.1007/978-3-030-99948-3_13

protocol to just under 21 millions [2], the only factors which could justify to attribute a certain value to Bitcoin. This influences its market stability and its usability as a reasonable currency apt to make investments. The Bitcoin's rate is subject to high volatility, not least as there exists no regulating body for it.

Using Bitcoins results in a new transaction, which contains a digital signature as proof and which relates to an earlier transaction. The new transaction transfers the value to one or several new owners, who are determined by their public (encryption) keys. This way transaction chains are formed reflecting the circulation of Bitcoins and the changes in their ownership.

Certain nodes in the P2P network, which are equipped with additional software and called miners, independently of each other assemble to blocks some 2000 received transaction data records comprising some 500 bytes each. When a miner has formed a new block, it tries to confirm the block by carrying out a Proof-of-Work. As depicted in Fig. 1, for this a variable in the block header called Nonce is modified in form of a linear search until the hash value of this header becomes smaller or equal to a given target value. In every execution of the search loop, the hash function SHA-256 is evaluated, which is very demanding in computing effort. The number of loop executions until reaching the termination condition cannot be predicted.

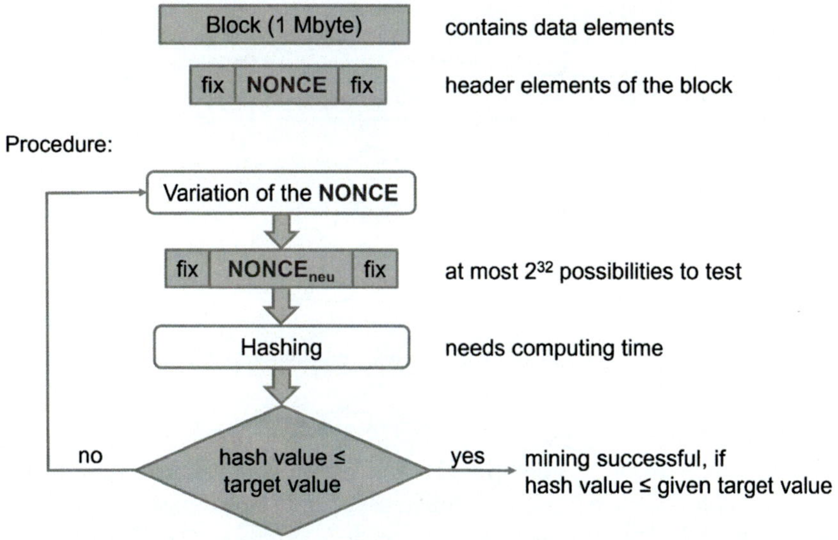

Fig. 1. Searching hash values by varying the Nonce in the block header.

The effort to execute the Proof-of-Work mechanism is regarded as a measure for the security and also for the acceptance of the Bitcoin system, because the higher the computing capacity required to build a new block is and, thus, the electrical power to be spent for this, the more difficult becomes an attack aiming to assemble—faster than the other nodes—a competing block with a transaction more favourable to the attacker.

Blocks are arranged in a linear chain, the so-called blockchain, cf. Fig. 2. Every block contains in its header the Merkle-Root hash value of the preceding block header. Thus, changes in a block appended earlier would automatically lead to inconsistencies in successive blocks, since the hash values recorded there would not be right anymore. Consequently, to verify the entire blockchain, only the linked block headers must the checked, but not all transactions in the blocks themselves, as the blocks' respective hash values are already recorded in their headers. Every network node provides its own copy of the blockchain, verifies independently of the others any new block received and, thus, any transaction, before it inserts the new block in its own local copy of the blockchain.

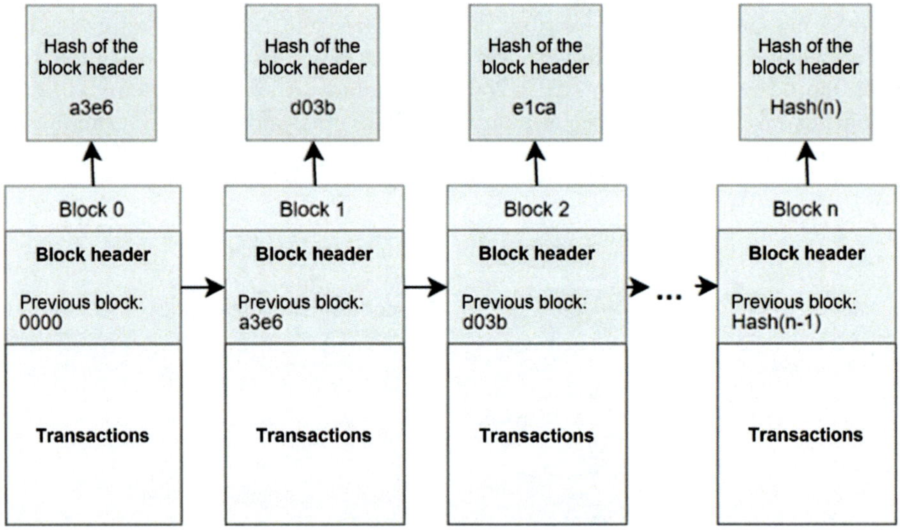

Fig. 2. Growing of the blockchain.

The Bitcoin protocol specifies to assemble a new block on average every 10 min. This block time is a compromise between a confirmation time to be as short as possible, and limiting the unnecessary effort required in case several competing local blockchains would arise. Such a situation occurs when two or more miners have built, almost at the same time, a new block each, and a node receives these different blocks (cf. below). Then, the latter node selects the longest blockchain as the currently valid one.

Bitcoin miners receive for the computing- and energy-intensive assembly of blocks remunerations in form of transaction fees and the so-called block reward. The latter is, respectively, realised by the first transaction contained in the block having no parent transaction, called coinbase transaction, and is connected with the creation of new Bitcoins. In addition to the Nonce, the time stamp in the block header and the Extra-Nonce field in the coinbase transaction may be varied to find a suitable hash value. This option, however, entails the re-calculation of the Merkle-Root hash value of all transactions contained in the block, since this hash value is part of the block header. Was the search for a hash value successful, i.e. was a valid block found, the miner propagates the block in the network.

The block reward, originally fixed as 50 BTC, is halved approximately every four years and, thus, limits the amount of Bitcoins that can be created to almost 21 millions. The miners receive as fees the differences between the input and the output values of the transactions included in the blocks. Therefore, miners prefer to include the transactions with the biggest differences. For the transaction fees a market price is reached, which depends on the number of transactions competing for inclusion in a block.

In order for a validly propagated block to be actually appended to the blockchain, first the receiving node must check whether all transactions contained are really valid, whether there is no double-spending, and whether the block header's hash value fulfills the requirements above. Nodes accept a block (achieve consensus) when they work on the successor block and use the hash value of the accepted block's header as previous hash value. The reward in BTC as mentioned above is credited for a successful hash value search. As valid blocks may be found almost simultaneously, however, it is possible that several branches of the blockchain are built. Sooner or later, though, one branch prevails as the longest one, cf. Fig. 3. The miners always aim to regard the longest chain as the correct one and, thus, to extend it.

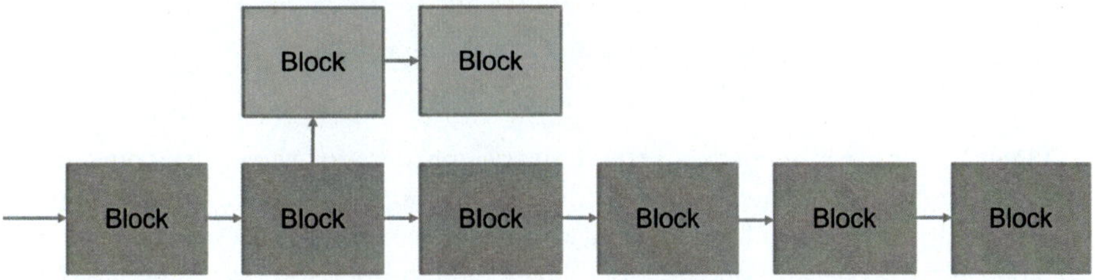

Fig. 3. Branches in the blockchain.

The blocks of the shorter branches are discarded. The transactions in the discarded blocks are, however, not lost as they occur in other valid blocks, too. For these reasons, a transaction is regarded as secure not before it appears in a block of the blockchain having at least five successor blocks, i.e. the blockchain has significantly grown since the transaction took place. As new blocks are built every 10 min on average, it may be acted on the assumption that a transaction can be considered as secured after approximately one hour.

Upon receiving a valid block, nodes must, as mentioned above, verify the transactions contained. In the course of this, it has to be taken into account that the Bitcoin value assigned to a Bitcoin address – with which exactly one public and one private key of the owner are associated – does not result from an electronic coin stored in a database, but derives from chains of digitally signed transactions. A received Bitcoin transaction T_n consists of the receiver's public key PKE_n, the number of Bitcoins to be transferred, a confirmation (signature) S_{n-1} of the previous owner (sender), and the latter's public key PKE_{n-1}. When generating a new transaction T_{n+1}, the sender signs with his private key PKD the hash value for the receiver's public key PKE_{n+1} and the previous transaction T_n. Figure 4 shows this relationship.

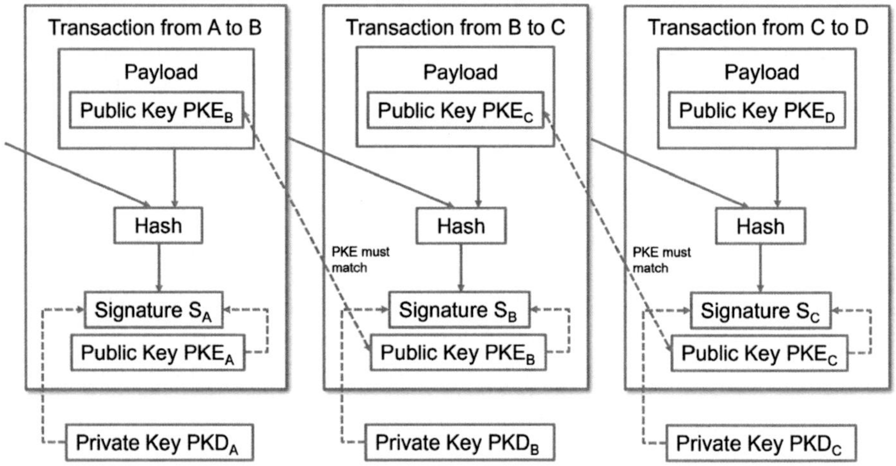

Fig. 4. Generation of transaction chains.

Thus, by repeatedly checking the digital signatures for the entire transaction chain, transactions are checked for validity. If all chain elements are in good order, then the recipient is, with high certainty, in possession of the Bitcoins transferred. The consensus procedure on the block level, as mentioned above, is necessary in order to detect double-spending. Hence, two kinds of chains are to be explicitly distinguished in this system: the chain of blocks containing transactions, and single transaction chains comprising transactions concatenated with each other.

The remarks above were to give an overview on the technical mechanisms in the Bitcoin system as well as on its features. Its fundamental idea is to secure payments by cryptographic techniques in this profoundly distributed system and, thus, to replace the classical trust model, which is based on financial institutions as trustworthy third parties. In the next section, it is to be scrutinised whether the Bitcoin system and the protocols connected with it can be considered as secure, capable of real-time operation, and as resource-adequate. In doing so, it is referred to the original publication of Satoshi Nakamoto, but not to the technical adjustments and extensions of the Bitcoin system made in recent years, which cope with these aspects more or less successfully indeed, but which do not constitute fundamental changes, and which let in effect the system become even more complex and inscrutable.

2 Discussion of Bitcoin's Concept

Bitcoins exist, as already mentioned, neither physically nor as balances on bank accounts, but are stored as chains of digital signatures on nodes of a Peer-to-Peer network ("We define an electronic coin as a chain of digital signatures" [1].). These nodes and their operators are pseudonym, i.e. their names are unknown, but hints to the nodes' identities may be obtained by observing the traffic within the network. Therefore, and because nodes may enter or leave the P2P network at will, the node operators elude any responsibilities, particularly legal ones such as liability. As nodes may arbitrarily leave the P2P network, it is possible that all Bitcoins disappear when all nodes leave the network, the network breaks down, or all active nodes fail.

The condition for Bitcoin transactions to be correctly executed, and for Bitcoins themselves not to be attackable is, according to Nakamoto, that "honest" nodes provide at least 51% of the computing power available in the network, because the longest chains generated by the most powerful group of nodes are the determining factors for concluding transactions. This requirement, however, cannot be complied with at any time, since node operators may deliberate why they should bear the procurement and operating costs of their nodes when the point is reached, where the profit gainable with generating new Bitcoins exceeds the expenses for the nodes. Thus, the concept enables criminals to capture the system very easily:

1. Operating nodes becomes uneconomical.
2. The "honest" operators (gradually) take their nodes out of service.
3. The majority in the P2P network devolves on criminal operators.

This way, Nakamoto achieves exactly the contrary to his claim: "We have proposed a system for electronic transactions without relying on trust" [1]. Indeed, Bitcoin users do not need to trust publicly known financial institutions, which are subject to official supervision of banking and which may be held liable, but need to trust unknown, often criminal node operators instead, and need to believe that a number of conditions is fulfilled at any time. The Bitcoin system is, thus, extremely insecure.

The same assessment is reached when considering how the system deals with using a single Bitcoin for multiple payments. To this end, a sender signs twice or several times the same transaction to different recipients. Such multiple spendings are to be prevented by majority votings of the nodes on which transaction with the same Bitcoin was the earliest one. For this purpose, the—as we shall see later, rather imprecise – transaction times are compared, Proof-of-Work is carried out, and blocks are checked. Nakamoto regards the multiple usage of Bitcoins as improbable in case attackers are slower than the honest nodes—but he cannot exclude it for sure.

In addition to the measure of probabilistically preventing multiple usage of Bitcoins, the following structural features of the Bitcoin system contribute to render its behaviour totally unpredictable. Namely, according to Nakamoto's design it is not required that the broadcasting of new transactions' data records reaches all nodes, and that best-effort routing is to be applied for this purpose. This is a further reason for the set of nodes forming the majority in the P2P network to change permanently. Furthermore, the transaction data are transmitted without encryption opening the floodgates to manipulations of any kind. Since the blockchain is constantly becoming longer by appending new transaction data, it may be shortened after some time. The Bitcoin system's unpredictability is aggravated by the lack of a specification when such truncations are to take place.

By applying normalisation rules on database schemes, relational datenbase systems ensure that no preventable data redundancies occur. Consequently, stored data remain consistent, and memory space is saved by preventing multiple data storage. Contrary to this, the Bitcoin blockchain is provided on the nodes of a P2P network in n-fold redundancy, and the consistency of its n versions is not cared for.

In comparison with classical computer science, which acted on the assumption of scarce resources and, thus, managed with virtual ones such as virtual memories or virtual machines, the concept of Bitcoin constitutes a paradigm shift. Its wasteful exploitation of energy, computing power, storage space, and bandwidth is exorbitant. If the aspect redundancy is considered, double or even triple bookkeeping is indeed feasible and common practice for centuries. In contrast, n-fold redundancy for rather large n does not yield any additional benefit and is, therefore, nonsense. Whereas just a few data transmissions suffice to process a financial transaction, in the Bitcoin system their number increases quadratically with the number of nodes.

The missing resource adequacy of Bitcoin does not only manifest itself in excessive use of resources, but also in the lack of resources like inadequate timing and temporal synchronisation as well as striking underestimation of storage requirements. Thus, as the amount of storage required annually, Nakamoto states completely unrealistic 4.2 MB by considering just the block headers and assuming six blocks per hour, only. However, the size of the transactions' data records is approximately half a kB, accumulating to some 53 GB between October 2020 and September 2021. Altogether, approximately 382 GB accrued from Bitcoin's launch until the end of 2021, a not insignificant amount of data, which has to be transferred to a node at least when putting it into operation.

Whereas a limited amount of payment units can be generated with minimal resource consumption, the generation of Bitcoins—the so-called mining—consumes prohibitively much electric current. In the year 2019 mining and transactions required 119 TWh power—8 TWh more than the Netherlands, an entire country with 17.5 million inhabitants [3]. The electric current necessary for a single Bitcoin operation is sufficient to light a 60 W bulp for 9 years and 187 days [4]. Most likely, the cost for electric power caused by a transaction thus exceeds the transaction's value in frequent cases.

In order to enable the determination of earliest payments or transactions, the corresponding data records must be endowed with time stamps. For this purpose, Nakamoto's concept envisages the presence of distributed time-stamping servers in the P2P network, but ignores that their clocks need to be synchronised, and that on a global scale and with very fine granularity. Technically, this is definitively possible [5, Chapter 14], e.g. by receiving the legal time Universal Time Co-ordinated (UTC) as disseminated by satellite navigation systems. On the other hand, organisatorically it is not ensured whether the nodes joining the P2P network are equipped with the required receiving devices, timers, and synchronisation capabilities. Four bytes are reserved in the block header to accommodate time stamps. With these 32 bits, the duration of one year can be resolved in steps of a little bit less than 10 ms. For a worldwide, frequently used payment system, however, a period under consideration of just one year is too short, and a temporal resolution of 10 ms too coarse by far. Thus, time stamps four bytes long prove totally insufficient.

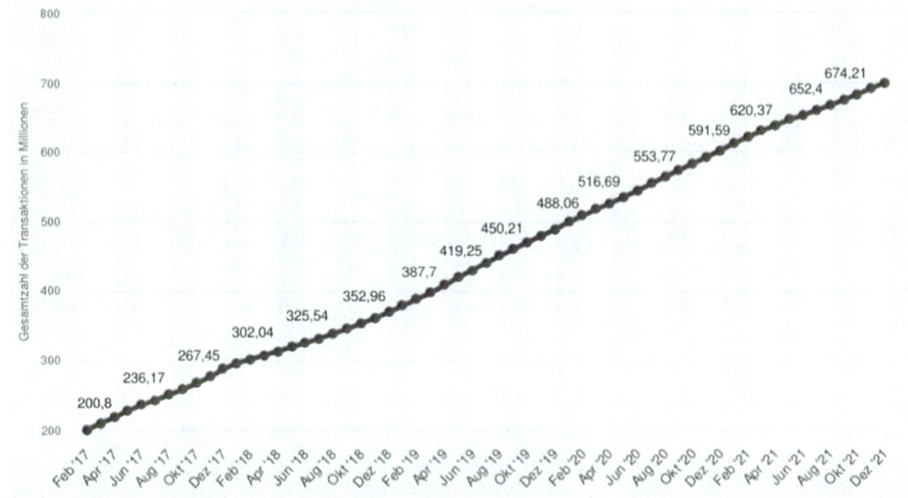

Fig. 5. Total number of worldwide Bitcoin transactions between February 2017 and December 2021 (in millions) [6].

With respect to response times in handling payments, payment systems must fulfill requirements ranging from low to extremely high. The latter occur, for instance, in high-frequency trading at stock exchanges and vary down to the nanosecond range. As stated above, however, Bitcoin transactions can only be regarded as completely executed and confirmed after waiting for ten minutes. Thus, Bitcoin is absolutely unsuitable in daily life as payment system, because no customer and no merchant would ever accept to wait ten minutes at a cash register for a payment's confirmation. When they were introduced, Bitcoins were presented as a "possibility for small casual transactions" [1], and happy people were shown paying at coffee dispensers with Bitcoins. It was concealed, however, that coffee becomes cool within ten minutes. Hence, one cannot talk about real-time capabilities in connection with Bitcoins.

Considering the number of transactions also reveals that Bitcoin is not suitable as a payment system ready around the clock for worldwide use. Figure 5 shows that from its introduction until September 2021 approximately 670 million Bitcoin operations were processed, some 100 million of them in the last twelve months [7]. On average this means a little bit more than just three transactions per second during the timeframe from October 2020 to September 2021. If just any eighth person on earth would induce only a single transaction per day, however, approximately 11575 transactions had to be processed per second. This number is a very conservative estimation considering the multitude of electronic payments—particularly small and smallest ones and specifically in Asia—carried out these days. These or even higher requirements would certainly overstrain the capabilities of the Bitcoin system by far.

3 Conclusion

After having presented the crypto currency Bitcoin and the processing of its transactions, its properties were discussed with reference to original citations of the defining document [1]. It turned out that the concept of the crypto currency is ill-conceived, and that its creator hiding behind the pseudonym Satoshi Nakamoto did obviously not understand it himself. The concept is far too complicated, for which reason the crypto currency is not suitable as a payment system usable worldwide. Particularly, the concept is evidence of complete ignorance of computer-based time management and of clock synchronisation in distributed systems. The system designed by Nakamoto works neither deterministically nor predictably—two properties being essential for a payment system. In addition, transactions with Bitcoins lack real-time capabilities and are neither secure nor resource-adequate. Indeed, the transactions do not need a (central) bank, but they are not anonymous, since the blockchain stores the complete money flows ("public history of transactions"). Thus, the ban of using Bitcoins as pronounced by the Chinese government for the prohibitively high electric power comsumption and to protect users from erratic value fluctuations was very reasonable.

Acknowledgement. The authors thank the three anonymous reviewers for their critical comments, which were predominantly negative, deeming the paper's content opinion, non-scholarly and unbalanced. The reviewers did not, however, provide fact-based arguments against the authors' "opinion", nor did they falsify any of the statements made. How could they actually endorse the authors' findings better?

References

1. Nakamoto, S.: Bitcoin: A Peer-to-Peer Electronic Cash System (2008). bitcoin.org
2. Meisner, H.: Bitcoins als Herausforderung in der Finanzsphäre. In: Lempp, J., Pitz, T., Sickmann, J. (eds.) Die Zukunft des Bargelds, pp. 89–102. Springer, Wiesbaden (2018). https://doi.org/10.1007/978-3-658-21720-4_6
3. Bocksch, R.: Bitcoin-Stromverbrauch - Bitcoin verbraucht mehr Strom als die Niederlande. Hamburg (2021). https://de.statista.com/infografik/18608/stromverbrauch-ausgewaehlter-laender-im-vergleich-mit-dem-des-bitcoins/. Accessed 31 Oct 2021
4. Brandt, M.: Digitalwährungen - Neue Kryptoprojekte bald so effizient wie Visa. Hamburg (2018). https://de.statista.com/infografik/13071/visa-und-krypto-coins-im-vergleich/. Accessed 31 Oct 2021
5. Halang, W.A., Konakovsky, R.M.: Sicherheitsgerichtete Echtzeitsysteme, 3rd edn. Springer, Berlin (2018). https://doi.org/10.1007/978-3-662-56369-4
6. Rabe, L.: Gesamtzahl aller Bitcoin-Transaktionen weltweit bis Dezember 2021 (2022). https://de.statista.com/statistik/daten/studie/315084/umfrage/gesamtzahl-aller-bitcoin-transaktionen-weltweit/. Accessed 14 Jan 2022
7. Statista. Bitcoin - Statista-Dossier zur virtuellen Währung Bitcoin. Hamburg (2021). https://de.statista.com/statistik/studie/id/22944/dokument/bitcoin-statista-dossier/. Accessed 31 Oct 2021

Journal Recommendation System for Author Using Thai and English Information from Manuscript

Nithirun Numnonda[1(✉)], Sapa Chanyachatchawan[2], and Nuengwong Tuaycharoen[1]

[1] Department of Computer Engineering, Faculty of Engineering, Chulalongkorn University, Bangkok, Thailand
6270142321@student.chula.ac.th, nuengwong.t@chula.ac.th

[2] National Electronics and Computer Technology Center (NECTEC), Bangkok, Thailand
sapa.cha@nectec.or.th

Abstract. There are thousands of academic journals in various fields of study. An article author must spend significant time searching and selecting a journal suitable for the article's content before submitting it to a journal for consideration. Since many articles are submitted to a journal at a time, it would take time for an editor to review, submit it to reviewers, and inform the results back to the author. If the article is rejected due to mismatched journal content, the author will spend more time to re-submit the article to another journal. Therefore, this research introduced a recommendation system to help the author choose an appropriate journal more effectively, based on TCI Thai Journals Online Database (ThaiJO). Data from Thai and English articles were used for analysis in this research. Our work involved studying the applied data, cleaning the data, and modeling, which included calculating the importance of text by Term Frequency - Inverse Document Frequency (TF-IDF), calculating similarity scores between articles and journals using Cosine Similarity and then ranking the scores to recommend the most suitable journal. This research experiment with modeling between a model from Thai data, a model from English data, and a model using both languages. The experiment shows that when we combine Thai and English keywords and abstract data, the accuracy in the form of hit rate is improved to 0.80650 from applying only English (0.78793) or Thai data (0.62888).

Keywords: Journal recommendation system · TF-IDF · Cosine similarity · Thai and English articles

1 Introduction

Since there are currently many academic journals in various fields of study, as a result, an author of an article must spend much time searching and selecting academic journals that are suitable for the content of the author's article before submitting the article to the journal editor for consideration. Scopus [1] indexes content from more than 25,000

P. Meesad et al. (Eds.): IC2IT 2022, LNNS 453, pp. 142–151, 2022.
https://doi.org/10.1007/978-3-030-99948-3_14

active titles and 7,000 publishers, reviewed and selected by independent committees. Users on Scopus have access to thousands of books, millions of author accounts, and 1.7 billion references. While Thailand has a platform, Thai Journals Online (ThaiJO) [2] is a collection of academic journals produced in Thailand in all disciplines, both in science/technology and in humanities/social sciences. ThaiJO collects more than 1,013 journals and has over 18,542 volumes with over 192,044 articles. Due to numerous journals from both Scopus and ThaiJO mentioned above, an author must spend time to select a journal to submit one's article to a journal for consideration. If the publication is rejected, it will take even more time. Coupled with the fact that the contents of some journals are in the same field of study, they have different concentrations, making it challenging for the article's author to choose the right journal for submitting.

Therefore, this research introduced a recommendation system to help the author choose an appropriate journal more effectively. The recommendation system is widely used today. Whether it is a web service, Streaming, E-Commerce, or even a social media website, all have recommendation systems to help screen content to service users to find things that most closely match their interests.

The recommendation system will take the information contained in the system for analysis—the more significant the amount of data in the system, the better the prediction results. The recommendation system uses a variety of algorithms and analytical methods. Many research papers use various information, whether text information from the entire article, article titles, abstracts, keywords of documents, or articles previously published by the article's author, including cited articles and referenced from articles. Likewise, there are methods of calculating the importance of the text of the word, such as calculating the importance of text by Term Frequency - Inverse Document Frequency (TF-IDF), calculating the significance of word by Term Frequency. For the data modeling, the are many techniques such as using the Clustering technique to determine to recommend the article, calculating similarity scores between articles using Cosine Similarity then ranking the scores to recommend the most relevant articles.

This article describes development of a recommendation system to help the system's analysis to suggest journals suitable for an article more effective for the author of the article. Therefore, we use the information of journals and various articles available in the TCI Thai Journals Online Database to analyze and create a recommendation system. Data from Thai and English articles were used for analysis, and comparative experiments were conducted between the models in each language. The Thai journal recommendation system will help promote Thai journals to facilitate and encourage more authors to publish in Thai journals.

2 Related Theories and Related Works

2.1 Related Theories

Text Cleansing. In the text information that we have obtained from the article, sometimes there are special characters or various marks that we do not want, for example, = +!?*/. These letters will affect word segmentation, including language processing. To improve the accuracy of the analysis, initial text cleaning [3] is required before model processing. There are many methods to deal with text, such as word segmentation, removing unnecessary or unimportant characters, and transforming the word to the original meaning to be processed more accurately.

Thai Word Segmentation. Word Segmentation [4] is the division of the text to find the boundaries of word units. This is the primary step of Natural Language Processing (NLP). In the Thai language, words are written next to each other without spaces except for the periodic spacing to allow the reader to take a break and understand the episodic meaning. This makes Thai more complicated to process than English, which has spaces to separate words. The PyThaiNLP library [5] is a Python library for natural language processing. It has basic capabilities for processing Thai language, such as segmentation and checking word spelling.

Calculating the Importance of Text. The Term Frequency - Inverse Document Frequency (TF-IDF) technique [6] is one of the methods for finding keywords in a document using the entire document's content. Word frequency alone does not indicate the importance of words in a document. For example, words may be abundant in the document, but they are unimportant, such as Stop Words. The Term Frequency - Inverse Document Frequency technique will get essential words from an article. It consists of two parts: Term Frequency (TF) and Inverse Document Frequency (IDF).

Since each document may have a different length, the Term Frequency (TF) calculation is usually divided by the number of words in the document. It can be calculated from the following formula.

$$\mathrm{TF}(\mathrm{term}, \mathrm{document}) = \frac{\mathrm{f}(\mathrm{term}, \mathrm{document})}{\sum_{\mathrm{term}' \ \mathrm{document}} \mathrm{f}(\mathrm{term}', \mathrm{document})} \tag{1}$$

Inverse Document Frequency (IDF) is used to measure the importance of words in all documents. If a word is found in multiple documents, that word becomes less important. It can be calculated from the following formula.

$$\mathrm{IDF}(\mathrm{term}, \mathrm{allDocument}) = \log \frac{N}{df(t)} \tag{2}$$

Finally, The Term Frequency - Inverse Document Frequency (TF-IDF) technique can be calculated by combining the Term Frequency (TF) and The Inverse Document Frequency (IDF) reciprocal values, calculated from the following formula.

$$\mathrm{TF\ IDF} = \mathrm{TF}(\mathrm{term}, \mathrm{document}) \times \mathrm{IDF}(\mathrm{term}, \mathrm{allDocument}) \tag{3}$$

Calculating Similarity Scores Between Documents. Several statistical methods measure similarity, including Jaccard Similarity and Cosine Similarity [7]. In this research, Cosine Similarity was used [8], calculated from the cosine angle between the two vectors. If the cosine value approaches one, then the two vectors are very similar. The cosine similarity measure can be calculated using the following formula.

$$Cosine(\overline{v}, \overline{w}) = \frac{\overline{v} \cdot \overline{w}}{|\overline{v}||\overline{w}|} = \frac{\sum_{i=0}^{n} v_i \times w_i}{\sqrt{\sum_{i=0}^{n} v_i^2}\sqrt{\sum_{i=0}^{n} w_i^2}} \tag{4}$$

K-Fold Cross-Validation. K-fold Cross-Validation [9] is a data validation by dividing the data into K parts, each with the same amount of data. Each part of the data is then used in a cycle to test the model's performance, which iterates until all K parts are used. The model's accuracy is obtained by averaging the accuracy of each cycle. Model validation with K-fold Cross-Validation was more reliable than a single trial Validation.

2.2 Related Work

Magara, Ojo, and Zuva [10] proposed a model for recommending academic papers based on the word frequencies in all texts of English articles. TF-IDF values were also calculated to indicate the importance of words. The documents were used to calculate the similarity value. In this study, several similarity measures were compared to adopt an appropriate similarity metric: Cosine Similarity, Euclidean distance, Jaccard coefficient, Pearson correlation coefficient. In the experiment, 220 articles on Artificial Intelligent (AI) from the Faculty of Computer Science at Ghent University were used with 70% for the training set and 30% for the testing set. The experiment compared the accuracy and efficiency of 3 algorithms, including Random Forest, Recursive Partitioning, and Boosted tree, to develop an academic paper recommendation system. As a result, the Recursive Partitioning algorithm was selected for the model development due to its high accuracy of 80.73% and processing time of 2.354628 s, while others have equal or better accuracy, with 20X slower.

Lee, Lee, and Kim [11] developed a model for recommending academic articles to the reader using the information from the articles that the reader wrote. The model used Web Crawler to extract data from two Academic article sites, including IEEE Xplore and The ACM Digital Library. It used the Title, Keyword, and Abstract sections to represent the word group to represent the article. To clean up the data, the researcher chose to reshape words to conform by eliminating letters ending "-ed", "-ly", and "-ing" from words and decided to remove the other 140 non-essential words that the researcher chose to cut off. To create the model, they defined user metrics based on the principle that the article owner will like their articles and put them as a preference level. Similarities between all articles and user articles were also calculated with the k-Nearest Neighbors (kNN) method. The research clustered articles into groups of articles in the same way as K-Means. The score was then calculated based on the distance between the target article and the center of the centroid. The evaluation included research from three groups: Machine learning, Database, and Human-Computer Interaction. By recommending ten articles out of a total of 10,386 articles for each group. The recommendation system suggested the correct group accounted for 89%.

Sugiyama and Kan [12] modeled for recommending academic articles using historical article data from each researcher, including articles citing the target article and articles cited by the target article. The first step is to create a user profile by dividing researchers into junior and senior researchers. A junior researcher will have only one article and no previously published articles. In comparison, senior researchers will have information on several previously published articles. Therefore, a junior researcher will use only the article information referenced by the target article. In contrast, the senior researcher's user profile uses the data of the researcher's previously published writings and the articles referenced by the target article. It is calculated from the Term Frequency instead of the Frequency - Inverse Document Frequency (TF-IDF). It then recommends articles by ranking their scores among the n highest ranks and those with the n highest rated values. In the experiment between article recommendations using only previously published articles, using cited articles, and being cited. The experiment results show that when reference article data was used in the model, it helped to recommend more accurate results than data from previously published articles only.

3 Methodology

In this research, the main action methods are Data Exploring, Data Preprocessing, and Data Modeling (Fig. 1).

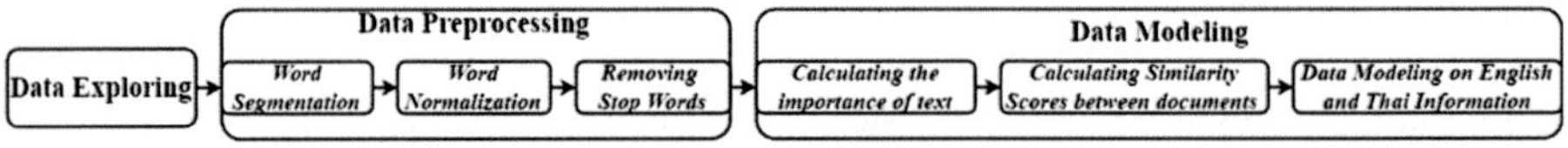

Fig. 1. Workflow of this research.

3.1 Data Exploring

The data used for analysis in this research are academic journal data obtained from Thai Journals Online (ThaiJO). This database system collects academic journals produced in Thailand in all disciplines. There are both Thai and English journals, some of which have information only in Thai, and some only have English information. The journals used for analysis in this research consisted of 3282 articles from 17 engineering and technology journals. Our criteria are to select journals only in the TCI Tier 1 group and Tier 2 group. Among 3282 articles, 1233 articles have both English and Thai information, but there are 2049 articles that have only English information.

3.2 Data Preprocessing

Based on the data obtained, we selected the information used in each article, including Article ID, Journal ID, Abstract Data, and Keyword Data. Each article's Thai and English data are processed in each language separately. The process for preparing information in both languages has a slightly different approach.

Word Segmentation. This is the process of dividing words from sentences. In this research, the process is as follows. In English, word segmentation is possible with spaces because English has a grammatical syntax where every word is spaced, so we use nltk, a natural language processing library for python, to process English text. However, words cannot be segmented with spaces in the Thai language. This makes word segment in the Thai language more complicated. Therefore, we use PyThaiNLP, a library for natural language processing in Thai.

Word Normalization. Both Thai and English have some words with similar meanings but different words according to sentence and context. In English, verbs are conjugated according to grammatical principles. For example, we can use ‘ran’ instead of ‘run’, ‘ate’ instead of ‘eat’, and ‘deleted’ instead of ‘delete’ when these sentences describe past events. Some words have the same root but are spelled differently because the word can be used as nouns, verbs, or prepositions such as prepare, preparation, prepared, etc. Therefore, it is necessary to restore these words to the word’s root at this stage. We used nltk to normalize English words. We also use PyThaiNLP for the Thai language data.

Removing Stop Words. Prepositions or connecting words are usually present in most sentences, and they are called Stop Words. Therefore, to find the keywords in each article, these words should be removed from the data to reduce the processing burden and increase the processing process’s accuracy. In English, the stop word is already defined in the nltk library, so functions from nltk can be used. As for the Thai language data, we use the data published on Stopwords ISO [13], which collects stop words in various languages, including Thai. The words in Stopwords ISO are identified and excluded from our data.

3.3 Data Modeling

Our system determines the importance of words in each article as keywords, and the keyword score was used to find similarities between each article and the journals to recommend an academic journal suitable for an article. We experimented with Thai and English data separately and then combined the Thai and English models for better efficiency. In both languages, there are procedures for calculating keywords using Term Frequency - Inverse Document Frequency (TF-IDF) and calculating the similarity of each article and journal using Cosine Similarity (Fig. 2).

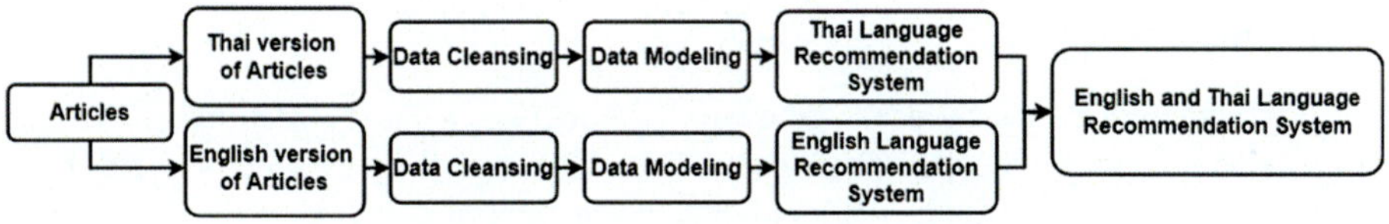

Fig. 2. Workflow of data modeling in this research.

Calculating the Importance of Text. An article in an academic journal will have a keyword list for each article, determined by the article’s author. The exact keywords are often presented in the same journals with the same field of study. However, the keywords that the article authors enter may not be accurate or cover all related keywords of the articles. Therefore, the Term Frequency - Inverse Document Frequency technique must be used to obtain better-valued keywords. In this research, the texts in the abstract and the keywords of each article in the same journal were combined and then calculated the Term Frequency - Inverse Document Frequency score.

Calculating Similarity Scores Between Documents. To find a suitable journal for an article, we assume that the suitable journal contains similar articles. Therefore, we calculate the Cosine Similarities values between the articles and all articles in each journal, rank the similarity scores between the articles and all articles in each journal, and recommend the journals with the top 5 highest score. Our system will recommend the top 5 appropriate journals because a qualified article may be accepted to be published in more than one journal with similar standards. For example, an article in a computer engineering journal of Publisher A may also be accepted for publication in a computer engineering journal of Publisher B if the author decides to submit it to the journal of Publisher B instead of the journal of Publisher A. The identity of the journal is based on the information from the articles in that journal, which uses several articles in the journal to develop the journal's identity with sufficient efficiency.

Data Modeling on English and Thai Information. We consider using both languages to improve the accuracy of the recommendation system. We adjust the similarity score weight using both language scores. If either language has a score of 0, the score of the other language is used instead. If both languages' scores have more than 0, we will adjust the score weight by a mean of both scores.

4 Experimental Result

4.1 Experimental Setup

We develop three language models based on data that we have, including English, Thai, and Both languages processed together (bilingual). The English, Thai, and bilingual models were used to compare the results. The experiments were conducted using keywords abstract data in each language.

4.2 Evaluation Metric

To validate the model, we choose K-fold Cross-Validation, a data validation by dividing the data into K parts. The test will be repeated until the entire K section has been reached. We chose K as 5 in this experiment, which is the most common choice [14] between k = 5 and k = 10 because these values were tested to be unaffected by excessive bias [15]. For each round, we use the Hit Rate to evaluate the model. The Hit Rate is calculated as the following: For an article in the test set, if the article's actual published journal is in the top 5 journals suggested by the model, we count as one Hit. Then, we divide the number of Hits by the total number of test data to obtain the Hit Rate. The closer the Hit Rate is to 1, the more accurate the model is. We aim for the model to have an accuracy of over 0.8 to be considered valid because our related work [10–12] claimed the accuracy of 0.8073, 0.89, and 0.739, which mean accuracy is 0.8121. We conclude that an accuracy of 0.8 is a reasonable target.

4.3 Experiment on English Information

In our experiment, when we model with only the keyword data and test with 5-fold cross-validation, the mean is 475.6 out of 656.4 articles, and the system suggests the original published journal among the top 5 appropriate journals. This accounts for a 0.72455 Hit Rate. However, when experimenting with abstract data, the results were lower, with a hit rate at 0.70445 or 462.4 articles out of a total of 656.4 articles. When we combine keywords and abstract data and adjust the weights of Cosine Similarity by assigning more values of the keyword than values of the abstract words, the hit rate is improved to 0.78793 or 517.2 articles of 656.4.

4.4 Experiment on Thai Information

Applying only the Thai language to the model provides lower accuracy than applying only English. The model with only Thai keyword data gives the correct result of 405 out of 656.4 articles, accounting for a hit rate of 0.61701. While experimenting with only Thai abstract data, the result is better with the hit rate at 0.62004 or 407 articles out of a total of 656.4. When we apply both Thai keywords and abstract data together and adjust the weight of Cosine Similarity same way as applying only English, the hit rate is improved to 0.62888 or 483 articles of 656.4.

4.5 Experiment on English and Thai Information

To apply both English and Thai Information, we create a function to determine the Cosine Similarity of English and Thai data and determine which Cosine Similarity we should use. The experimental result of using both languages shows higher accuracy than applying only either English or Thai. The result of the model with only keyword data is a 0.74617 Hit Rate or 498.8 articles of 656.4. When experimenting with only abstract data, the result is 473.6 articles out of 656.4, accounting for a hit rate of 0.70445. Finally, when we combine keywords and abstract data, the hit rate is improved to 0.80650 or 529.4 articles of 656.4 (Table 1).

Table 1. Compare each language model experiment.

Language	Data on section	Correct suggest article	Hit rate accuracy
English information	Keywords	475.6 articles of 656.4	0.72455
	Abstract	462.4 articles of 656.4	0.70445
	Keywords and abstract	517.2 articles of 656.4	0.78793
Thai information	Keywords	405.0 articles of 656.4	0.61701
	Abstract	407.0 articles of 656.4	0.62004
	Keywords and abstract	483.0 articles of 656.4	0.62888
English and Thai information	Keywords	489.8 articles of 656.4	0.74617
	Abstract	473.6 articles of 656.4	0.70445
	Keywords and abstract	529.4 articles of 656.4	0.80650

5 Conclusions and Future Works

This research proposes a recommendation system to help an author choose an appropriate journal more effectively using information available on the TCI Thai Journals Online Database (tci-thaijo.org), which uses data from engineering and technology journals by using both English and Thai data on keyword and abstract sections.

In this research, we develop a model based on journal data to compare keyword data, abstract data, and the combination of the keyword and abstract, and we have experimented with applying only English data, only Thai data, and combined Thai and English data models. From the experiment, we have presented guidelines for developing a model for recommending academic journals suitable for the authors of the articles. Our model using only English has an accuracy in the form of a hit rate of 0.78793, and our model using Thai only has the hit rate of 0.62888, But when we combine the two languages, the hit rate is 0.80650, which is higher than both the English-only model and the Thai-only model.

There were obstacles in data selection in this research because we wanted to recommend suitable journals, but some journals accept articles in too many related fields. Thus, it was necessary to select suitable journals for modeling.

For future works, we plan to apply the journal recommendation system with users and measure their satisfaction and confidence. We will consider various sets of journals with a large amount of data in the development of a more accurate model.

References

1. Scopus. https://www.elsevier.com/solutions/scopus/how-scopus-works/content. Accessed 10 Oct 2021
2. ThaiJo. https://www.tci-thaijo.org/. Accessed 21 Oct 2021
3. Haddi, E., Liu, X., Shi, Y.: The role of text pre-processing in sentiment analysis. Procedia Comput. Sci. **17**, 26–32 (2013)
4. Sornlertlamvanich, V.: Word segmentation for Thai in machine translation system. Mach. Transl. NECTEC, 556–561 (1993)
5. PyThaiNLP. https://pythainlp.github.io/docs/2.3/. Accessed 12 Aug 2021
6. Robertson, S.: Understanding inverse document frequency: on theoretical arguments for IDF. J. Doc. **60**(5), 503–520 (2004). https://doi.org/10.1108/00220410410560582
7. Mizzaro, S., Pavan, M., Scagnetto, I.: Content-based similarity of Twitter users. In: Hanbury, A., Kazai, G., Rauber, A., Fuhr, N. (eds.) ECIR 2015. LNCS, vol. 9022, pp. 507–512. Springer, Cham (2015). https://doi.org/10.1007/978-3-319-16354-3_56
8. Li, B., Han, L.: Distance weighted cosine similarity measure for text classification. In: Yin, H., et al. (eds.) IDEAL 2013. LNCS, vol. 8206, pp. 611–618. Springer, Heidelberg (2013). https://doi.org/10.1007/978-3-642-41278-3_74
9. Moreno-Torres, J.G., Saez, J.A., Herrera, F.: Study on the impact of partition-induced dataset shift on k-fold cross-validation. IEEE Trans. Neural Netw. Learn. Syst. **23**(8), 1304–1312 (2012). https://doi.org/10.1109/TNNLS.2012.2199516
10. Magara, M.B., Ojo, S.O., Zuva, T.: A comparative analysis of text similarity measures and algorithms in research paper recommender systems. In: 2018 Conference on Information Communications Technology and Society (ICTAS), pp. 1–5 (2018)
11. Lee, J., Lee, K., Kim, J.: Personalized academic research paper recommendation system. arXiv:1304.5457 (2013)

12. Sugiyama, K., Kan, M.Y.: Scholarly paper recommendation via user's recent research interests. In: Proceedings of the 10th Annual Joint Conference on Digital Libraries, Gold Coast, Queensland, Australia, pp. 29–38. Association for Computing Machinery (2010)
13. Stopwords ISO. https://github.com/stopwords-iso/stopwords-iso. Accessed 4 Aug 2021
14. Kuhn, M., Johnson, K.: Over-fitting and model tuning. In: Kuhn, M., Johnson, K. (eds.) Applied Predictive Modeling, pp. 61–92. Springer, New York (2013). https://doi.org/10.1007/978-1-4614-6849-3_4
15. James, G., Witten, D., Hastie, T., Tibshirani, R.: Resampling methods. In: James, G., Witten, D., Hastie, T., Tibshirani, R. (eds.) An Introduction to Statistical Learning: with Applications in R, pp. 175–201. Springer, New York (2013). https://doi.org/10.1007/978-1-4614-7138-7_5

Efficiency Optimization for Operational Performance in Green Supply Chain Sourcing Using Data Envelopment Analysis: An Empirical Study

Kasin Ransikarbum[1](✉), Chewaphorn Chaiyaphan[1], Siwaporn Suksee[2], and Sombat Sinthuchao[1]

[1] Ubon Ratchathani University, Ubon Ratchathani, Thailand
{kasin.r,chewaphorn.c}@ubu.ac.th
[2] Surindra Rajabhat University, Surin, Thailand

Abstract. This research aims to analyze the Technical Efficiency (TE) of the Participatory Guaranteed System (PGS)-certified farmers in the green supply chain network at Ubon Ratchathani Province. Linear optimization programming-based Data Envelopment Analysis (DEA) technique is used to evaluate selective samples of farmers collected from purposive sampling technique. In particular, data are initially collected from 61 farmers using questionnaire to gather 8 input criteria (i.e., IC1) planting area, IC2) number of the vegetable types, IC3) cost of fertilizers or soil, IC4) cost of seed, IC5) labor cost, IC6) transportation cost, IC7) cost of fuel used in agricultural machinery, and IC8) machinery maintenance costs) and 2 output criteria (i.e., OC1) income from sales of crop inputs, such as compost, planting soil, seeds, and OC2) income from sales of produced crops), respectively. Then, participatory farmers' operational performances are analyzed based on an overall efficiency for comparative study, in which about 34% of all farmers are found to be overall efficient. Next, an assess based on the cluster of district location for 8 districts shows that there are 2 districts operated with full efficiency (i.e., 1.00), 4 districts operated with high efficiency (i.e., 0.801–0.999), and 2 districts with medium efficiency (i.e., 0.601–0.800). Our initial results are to be used to guide the development of decision-support logistical system of location-routing optimization in the green supply chain, which is an on-going study of our research.

Keywords: Efficiency optimization · Data envelopment analysis · Operational performance · Participatory guaranteed system · Green supply chain

1 Introduction and Motivation

Supply Chain Management (SCM) is considered a management of a network system consisting of various nodes and arcs, where nodes are related to different locations from downstream to upstream (i.e., suppliers, production activities, warehouse, and end customers) and arcs are concerned with links to connect nodes (i.e., transporting product

P. Meesad et al. (Eds.): IC2IT 2022, LNNS 453, pp. 152–162, 2022.
https://doi.org/10.1007/978-3-030-99948-3_15

flows) [1]. Since its development, SCM has been intensively used in a number of applications with a main aim to optimize key performance index of concerned stakeholders [2, 3]. Green supply chain (GSC) is an emerging concept based on the SCM discipline to advance environmental sustainability with diverse green initiatives. GSC has lately intrigued a number of practitioners and researchers due to its capability to deal with not only the economic perspective, but also environmental and social issues. The terms GSC management (GSCM) is thus developed by integrating ecofriendly management and SCM to enhance environmental sustainability with various green practices related to sourcing, manufacturing, distribution, and transportation [4]. We note that a number of studies have been proposed relevant to the future research directions related to GSCM, which can be found from, for example, Maditati et al. [5] and Kaliani Sundram et al. [6].

In this study, we evaluate the sourcing decisions by assessing the efficiency optimization of the Participatory Guaranteed System (PGS)-certified farmers in the GSC context using a case study at Ubon Ratchathani Province, Thailand. We note that PGS certified crops are similar to the organic crops, which refers to plants that farmers rely on natural substances and biologically based farming methods without using prohibited substances to support the food-safety concern. Regardless, while organics certification requires expensive third-party audits, the PGS is developed based on a collaborative approach that involves farmers and other local stakeholders for verifying the authenticity of organic source. Thus, the PGS helps to make organic-farming certification possible for small-scale farmers, which are the main suppliers of the chain in our case study. [7]. Then, Linear Programming (LP)-based Data Envelopment Analysis (DEA) technique is used to assess selective samples of PGS-certified farmers, the upstream of the evaluated food-safety system in the GSC context, collected from purposive sampling technique. We note that this the current study is the first phase of our integrative analysis of the integrated location-routing model of the green supply chain, in which the efficient sourcing farmers are to be used as input dataset of the developed model. We further illustrate the GSCM in the context of food-safety system as shown in Fig. 1.

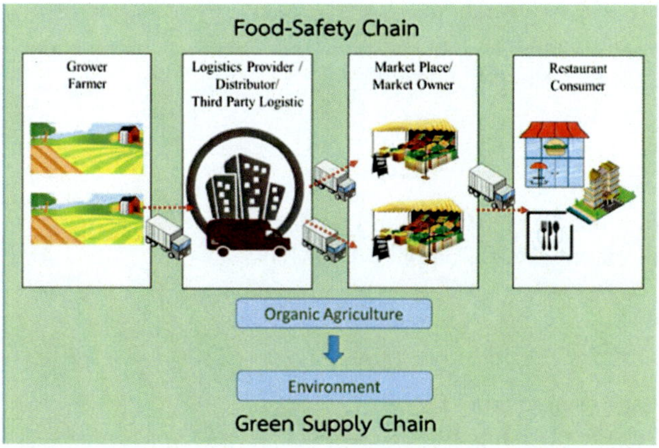

Fig. 1. A system of GSCM within the context of food-safety chain.

2 Data Envelopment Analysis (DEA)

We next discuss the Data Envelopment Analysis (DEA) technique implemented in this study to optimize the relative efficiency of Decision Making Units (DMUs) of interest. DEA is a multi-factor productivity analysis model that compares each variable with the best performing one, which has been initially used for both the production theory in economics and for benchmarking in operations management [8–10]. With respect to DEA development, DEA is considered a type of Multi-Criteria Decision Analysis (MCDM) techniques that explicitly considers multiple criteria in a decision-making environment and has been used to support decision making involving planning problems that a unique optimal solution does not exist and typically involves decision-maker's preferences [11–14]. Advantages of DEA include the capability to handle multiple inputs and outputs, where the sources of inefficiency can be analyzed and quantified for every evaluated unit. In particular, DEA can be extended in the form of LP methodology to measure relative efficiency of multiple DMUs with multiple input and output criteria/objectives. When the DEA model is solved for optimal solution, a particular DMU will be considered efficient if the objective score is one (i.e., 100%), whereas scores that are lesser than one imply relative inefficiency. It is also possible that more than one alternatives are found to be efficient.

We next illustrate the CCR model of DEA commonly used in the literature, which was early developed and named after the three researchers (Charnes, Cooper and Rhodes [15]) to measure the relative efficiency called Technical Efficiency (TE). Mathematically, the relative efficiency of a particular DMU can be obtained by solving the following model (Eqs. (1)–(4)), where the objective function is to maximize the ratio of the weighted sum of the outputs to the weighted sum of the inputs. The non-linear form of the model can be linearly transformed to the LP model as shown in Eqs. (5)–(9), respectively.

2.1 Sets

I: Set of input criteria, indexed by i
J: Set of output criteria, indexed by j
K: Set of DMUs (i.e., farmers), indexed by k.

2.2 Parameters

x_{i,k_0}: Amount of input data for input i of DMU k
y_{j,k_0}: Amount of output data for output j of DMU k.

2.3 Decision Variables

U_i: The weight assigned to input i
V_j: The weight assigned to output j.

2.4 DEA Mathematical Model

$$\text{Maximize Z:} \quad \frac{\sum_{j \in J} y_{j,k_0} V_j}{\sum_{i \in I} x_{i,k_0} U_i} \tag{1}$$

$$\text{Subject to} \quad \frac{\sum_{j \in J} y_{j,k} V_j}{\sum_{i \in I} x_{i,k} U_i} \leq 1; \forall k \in K \tag{2}$$

$$U_i \geq 0; \forall i \in I \tag{3}$$

$$V_j \geq 0; \forall j \in J \tag{4}$$

2.5 DEA Mathematical Model (LP)

$$\text{Maximize Z:} \quad \sum_{j \in J} y_{j,k_0} V_j \tag{5}$$

$$\text{Subject to} \quad \sum_{i \in I} x_{i,k_0} U_i = 1 \tag{6}$$

$$\sum_{j \in J} y_{j,k} V_j - \sum_{i \in I} x_{i,k} U_i \leq 0; \forall k \in K \tag{7}$$

$$U_i \geq 0; \forall i \in I \tag{8}$$

$$V_j \geq 0; \forall j \in J \tag{9}$$

3 Case Study of Operational Efficient Sourcing

3.1 Research Design

We next discuss the research design of the case study. Initially, a discussion session with management team of the PGS group is organized to purposively select representatives of farmers to evaluate operational performance. Thus, 61 representative farmers are chosen, in which a session of focus group with questionnaires is used to collect data from these farmers during the fourth quarter of 2021. Overall profile data as well as operational information are thus collected. Then, 8 input-criteria (IC) data are examined, which are IC1) planting area (Unit: Rai), IC2) number of the vegetable types (Unit: Type), IC3) cost of fertilizers or soil (Unit: Baht), IC4) cost of seed (Unit: Baht), IC5) labor cost (Unit: Baht), IC6) transportation cost (Unit: Baht), IC7) cost of fuel used in agricultural machinery (Unit: Baht), and IC8) machinery maintenance cost (Unit: Baht). In addition, 2 output criteria are further assessed, which are OC1) income from sales of production inputs such as compost, planting soil, seeds (Unit: Baht), and OC2) income from sales of produced crops (Unit: Baht), respectively. Then, participatory farmers' operational performances are analyzed based on their origin of district location for comparative evaluation of 8 districts (D), which are D1) Det Udom District, D2) Meung Ubon Ratchathani District, D3) Phibun Mangsahan District, D4) Warin Chamrap District, D5) Sawangveerawong District, D6) Samrong District, D7) Tansum District, and D8) Nayia District. In particular, there are 5, 12, 5, 13, 10, 7, 6, and 3 DMUs (i.e., farmers) associated with these districts.

The general location profile of the collected samples is illustrated using QGIS in Fig. 2. Clearly, farmers are found to be dispersedly located in a number of locations of their farm lands. In addition, Table 1 illustrates general profile of the input and output criteria collected from the farmer samples. We illustrate mean, median, and standard deviation (SD) associate with each criterion as shown in the table.

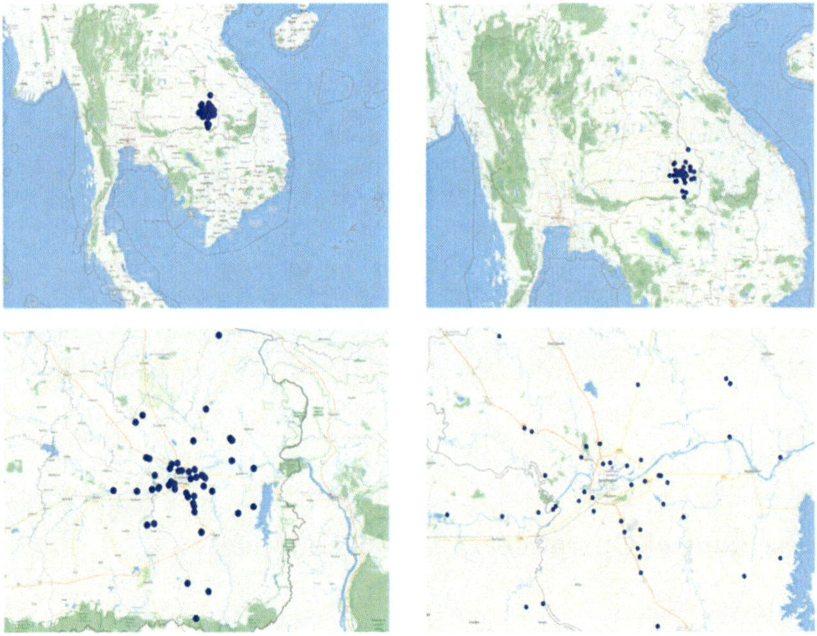

Fig. 2. General location profile of the evaluated farmers.

Table 1. General data profile of the input and output criteria from evaluated farmers.

Criteria	Description	Mean	Median	S.D
IC1	Planting area (Rai)	7.82	5.00	8.36
IC2	Number of crop types (Types)	6.92	6.00	3.58
IC3	Cost of fertilizers or soil (Baht)	10,266	6,000	10,810
IC4	Cost of seed (Baht)	2,876	1,493	4,316

(continued)

Table 1. *(continued)*

Criteria	Description	Mean	Median	S.D
IC5	Cost of labor (Baht)	58,370	22,712	85,023
IC6	Cost of transportation (Baht)	8,944	4,800	9,919
IC7	Cost of fuel machinery (Baht)	4,776	3,000	5,720
IC8	Cost of maintenance (Baht)	4,711	2,858	5,875
OC1	Income from sales of crop (Baht)	19,179	2,000	52,951
OC2	Income from sales of produced crops (Baht)	109,727	66,900	131,631

3.2 Analyzed Results and Discussion

We now illustrate the results from the DEA analysis based on multiple input and output to obtain operational efficiency of each sourcing location of farmers. In this study, we model the LP model of DEA in LINDO software [16]. The computational time is found to be essentially negligible in seconds. Initially, all the 61 farmers are assessed and ranked based on the overall efficiency as shown in Table 2. DEA analysis is helpful to provide insights with regard to relative efficiency of each alternative whether it is operated efficiently when multiple input criteria and output criteria are simultaneously considered. We note, however, that it is possible that more than one DMU may be found efficiently. In particular, our initial results suggest that 21 out of 61 farmers (i.e., approximately 34%) are efficient and are to be further recommended for the sourcing locations and can be acted as the representatives of the distributors among sourcing peers. Figure 3 also illustrates the Pareto chart of all analyzed results.

Table 2. Analyzed operational efficiency of overall farmers using DEA.

District	DMUs	TE	District	DMUs	TE
D1	DMU1	1.000	D7	DMU1	0.480
D1	DMU3	1.000	D3	DMU2	0.439
D1	DMU4	1.000	D6	DMU3	0.399
D1	DMU5	1.000	D4	DMU7	0.387
D2	DMU1	1.000	D7	DMU2	0.385
D2	DMU9	1.000	D5	DMU1	0.372
D2	DMU11	1.000	D3	DMU5	0.362
D2	DMU12	1.000	D7	DMU3	0.355
D7	DMU5	1.000	D6	DMU2	0.351
D8	DMU1	1.000	D4	DMU9	0.349
D8	DMU3	1.000	D4	DMU2	0.341

(continued)

Table 2. (*continued*)

District	DMUs	TE	District	DMUs	TE
D4	DMU4	1.000	D5	DMU2	0.340
D4	DMU5	1.000	D7	DMU6	0.289
D4	DMU6	1.000	D3	DMU1	0.288
D4	DMU8	1.000	D2	DMU8	0.286
D4	DMU10	1.000	D2	DMU6	0.269
D4	DMU12	1.000	D4	DMU1	0.269
D4	DMU13	1.000	D6	DMU4	0.268
D5	DMU4	1.000	D4	DMU3	0.242
D5	DMU8	1.000	D2	DMU2	0.236
D5	DMU9	1.000	D1	DMU2	0.200
D2	DMU3	0.971	D6	DMU5	0.194
D8	DMU2	0.944	D6	DMU7	0.188
D3	DMU3	0.919	D4	DMU11	0.186
D5	DMU7	0.862	D5	DMU6	0.142
D2	DMU5	0.733	D3	DMU4	0.138
D6	DMU6	0.699	D7	DMU4	0.110
D2	DMU10	0.656	D6	DMU1	0.095
D2	DMU4	0.577	D5	DMU10	0.094
D5	DMU3	0.504	D2	DMU7	0.082
D5	DMU5	0.497			

Next, we further examine these farmers based on locational clusters, in which each farmer is evaluated according to the locational area of 8 districts as shown in Table 3. Our results suggest that farmers that are located in the suburb area may have some more or less difficulties in managing high-cost resources than those in the rural area. Additionally, output acquisitions are also challenging in some other sourcing areas. Figure 4 also illustrates the results visually using bar chart diagram based on each cluster of farmers.

Table 3. Analyzed operational efficiency of clusters of farmer districts using DEA.

District	DMUs	TE	District	DMUs	TE	District	DMUs	TE
D1	DMU1	1.000	D2	DMU1	1.000	D3	DMU1	1.000
D1	DMU2	0.781	D2	DMU2	0.311	D3	DMU2	1.000
D1	DMU3	1.000	D2	DMU3	1.000	D3	DMU3	1.000
D1	DMU4	1.000	D2	DMU4	1.000	D3	DMU4	0.714

(*continued*)

Table 3. (*continued*)

District	DMUs	TE	District	DMUs	TE	District	DMUs	TE
D1	DMU5	1.000	D2	DMU5	1.000	D3	DMU5	1.000
	Mean	*0.962*	D2	DMU6	0.364		*Mean*	*0.943*
			D2	DMU7	0.162			
			D2	DMU8	0.563			
			D2	DMU9	1.000			
			D2	DMU10	1.000			
			D2	DMU11	1.000			
			D2	DMU12	1.000			
				Mean	*0.782*			
D4	DMU1	0.613	D5	DMU1	1.000	D6	DMU1	1.000
D4	DMU2	0.693	D5	DMU2	0.822	D6	DMU2	1.000
D4	DMU3	0.331	D5	DMU3	0.651	D6	DMU3	1.000
D4	DMU4	1.000	D5	DMU4	1.000	D6	DMU4	1.000
D4	DMU5	1.000	D5	DMU5	1.000	D6	DMU5	0.651
D4	DMU6	1.000	D5	DMU6	0.675	D6	DMU6	1.000
D4	DMU7	0.980	D5	DMU7	1.000	D6	DMU7	0.922
D4	DMU8	1.000	D5	DMU8	1.000		*Mean*	*0.941*
D4	DMU9	0.482	D5	DMU9	1.000			
D4	DMU10	1.000	D5	DMU10	0.321			
D4	DMU11	0.241		*Mean*	*0.850*			
D4	DMU12	1.000						
D4	DMU13	1.000						
	Mean	*0.791*						
D7	DMU1	1.000	D8	DMU1	1.000			
D7	DMU2	1.000	D8	DMU2	1.000			
D7	DMU3	1.000	D8	DMU3	1.000			
D7	DMU4	1.000		*Mean*	*1.000*			
D7	DMU5	1.000						
D7	DMU6	1.000						
	Mean	*1.000*						

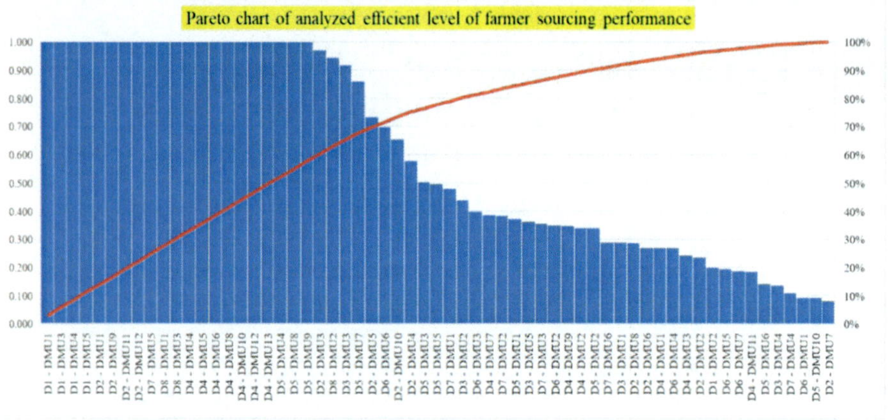

Fig. 3. Pareto chart of overall DEA score for all farmers.

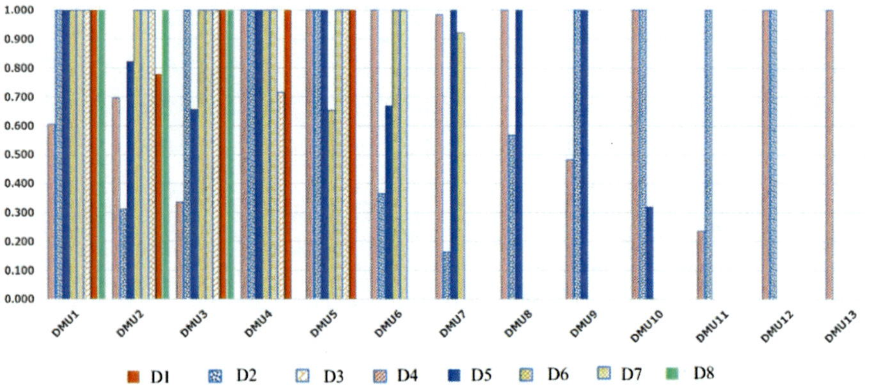

Fig. 4. Visualized DEA score based on the cluster analysis of 8 districts.

3.3 Discussion

Based on the above analysis of the operational performance of each farmer as a possible sourcing decision of the upstream of GSC, our results provide data for efficiently optimized sourcing DMU based on the DEA technique. An initial analysis for overall efficiency for all DMUs for a comparative evaluation of all DMUs suggests that about 34% of all farmers are operated efficiently when comparing to other DMUs. In addition, when farmers are analyzed based on each cluster of locational area to see a particular performance of DMUs in each cluster, our results show that there are 2 districts operated with full efficiency (i.e., 1.00), 4 districts operated with high efficiency (i.e., 0.801–0.999), and 2 districts with medium efficiency (i.e., 0.601–0.800).

It is important to note that the clustering analysis using DEA provides an evaluation of relative efficiency for each particular cluster, which differs from the concept of absolute efficiency [17–19]. That is, DEA technique is capable of finding how well a particular DMU is performing in comparison to its peers in an analysis but not with the ideal theoretical scenario. Accordingly, it follows in our analysis that when all the DMUs in a particular cluster are assessed, other DMUS in different clusters will be ignored. Thus, all the farmers that are found to be efficiently operated in a particular cluster implies that they are all operated competitively and efficiently in their own cluster.

4 Conclusion and Future Research Directions

In this study, an optimization analysis of the technical efficiency of the sourcing decision of farmers in the upstream process of the green supply chain network at Ubon Ratchathani province is conducted. Initially, we have coordinated with relevant parties involving in the food-safety system in the chain, especially with the staffs taking in charge for the local-level quality assurance system in the area called Participatory Guarantee System (PGS) to collect and assess operational performance data from agreed sample groups of possible farmers for sourcing decision. Both input and output data concerning operational efficiencies are then collected from agricultural farmers and are analyzed. Accordingly, we have successfully analyzed and evaluated operational efficiency of selective farmer samples involving in the research. We note that this is the initial study of our on-going work to examine the development of decision-support logistical system of location-routing optimization in the green supply chain. That is, agricultural distribution location and routing analyses are to be further conducted by considering both for the bottom-up aspect (i.e., farmer perspective) and for the top-down aspect (i.e., city development perspective) in our future research.

Acknowledgement. This research was supported by the Office of National Higher Education Science Research and Innovation Policy Council (PMUA) under project "*Development of Logistics Management System and Technology Transfer for Food-Safety in Green Supply Chain in Ubonratchathani*" (Grant number A15F640061). We note that the opinions expressed are those of the authors and do not necessarily reflect the views of the funding agencies.

References

1. Khan, S.A.R., Yu, Z., Golpira, H., Sharif, A., Mardani, A.: A state-of-the-art review and meta-analysis on sustainable supply chain management: future research directions. J. Clean. Prod. **278**, 123357 (2021)
2. Fritz, M.M., Rauter, R., Baumgartner, R.J., Dentchev, N.: A supply chain perspective of stakeholder identification as a tool for responsible policy and decision-making. Environ. Sci. Policy **81**, 63–76 (2018)
3. Wattanasaeng, N., Ransikarbum, K.: Model and analysis of economic-and risk-based objective optimization problem for plant location within industrial estates using epsilon-constraint algorithms. Computation **9**(4), 46, 1–22 (2021)
4. Ransikarbum, K., Chaiyaphan, C., Pataratanased, R.: Analysis of logistical aspect of food-safety system in the green supply chain using vehicle routing problem model. In: 2021 Research, Invention, and Innovation Congress: Innovation Electricals and Electronics (RI2C), Bangkok, pp. 48-53. IEEE (2021)

5. Maditati, D.R., Munim, Z.H., Schramm, H.J., Kummer, S.: A review of green supply chain management: from bibliometric analysis to a conceptual framework and future research directions. Resour. Conserv. Recycl. **139**, 150–162 (2018)
6. Kaliani Sundram, V.P., Rajagopal, P., Bahrin, A.S., Subramaniam, G.: The role of supply chain integration on green practices and performance in a supply chain context. A conceptual approach to future research. Int. J. Supply Chain Manag. **7**(1), 95–104 (2018)
7. Montefrio, M.J.F., Johnson, A.T.: Politics in participatory guarantee systems for organic food production. J. Rural Stud. **65**, 1–11 (2019)
8. Ransikarbum, K., Pitakaso, R., Kim, N., Ma, J.: Multicriteria decision analysis framework for part orientation analysis in additive manufacturing. J. Comput. Des. Eng. **8**(4), 1141–1157 (2021)
9. Ransikarbum, K., Pitakaso, R.: Relative efficiency analysis of biomass agricultural plants using data envelopment analysis. In: E3S Web of Conferences, no. 302. EDP Sciences (2021)
10. Ransikarbum, K., Pitakaso, R., Kim, N.: Evaluation of assembly part build orientation in additive manufacturing environment using data envelopment analysis. In: MATEC Web of Conferences, Vol. 293, p. 02002. EDP Sciences (2019)
11. Puchongkawarin, C., Ransikarbum, K.: An integrative decision support system for improving tourism logistics and public transportation in Thailand. Tour. Plann. Dev. **18**(6), 614–629 (2021)
12. Chanthakhot, W., Ransikarbum, K.: Integrated IEW-TOPSIS and fire dynamics simulation for agent-based evacuation modeling in industrial safety. Safety **7**(2), 47 (2021)
13. Ransikarbum, K., Khamhong, P.: Integrated fuzzy analytic hierarchy process and technique for order of preference by similarity to ideal solution for additive manufacturing printer selection. J. Mater. Eng. Perform. **30**(9), 6481–6492 (2021). https://doi.org/10.1007/s11665-021-05816-y
14. Ransikarbum, K., Leksomboon, R.: Analytic hierarchy process approach for healthcare educational media selection: additive manufacturing inspired study. In: 2021 IEEE 8th International Conference on Industrial Engineering and Applications (ICIEA), pp. 154–158. IEEE (2021)
15. Emrouznejad, A., Banker, R.D., Neralic, L.: Advances in data envelopment analysis: celebrating the 40th anniversary of DEA and the 100th anniversary of Professor Abraham Charnes' birthday. Eur. J. Oper. Res. **278**(2), 365–367 (2019)
16. LINDO. https://www.lindo.com/. Accessed 10 Oct 2020
17. Dyson, R.G., Allen, R., Camanho, A.S., Podinovski, V.V., Sarrico, C.S., Shale, E.A.: Pitfalls and protocols in DEA. Eur. J. Oper. Res. **132**(2), 245–259 (2001)
18. Ransikarbum, K.: Decision Analysis for Engineering. Chulalongkorn University Printing House, Bangkok (2021). (in Thai)
19. Pineda, P.J.G., Liou, J.J., Hsu, C.C., Chuang, Y.C.: An integrated MCDM model for improving airline operational and financial performance. J. Air Transp. Manag. **68**, 103–117 (2018)

Automatic Question and Answer Generation from Thai Sentences

Saranlita Chotirat[1(✉)] and Phayung Meesad[2]

[1] Information Technology Program, King Mongkut's University of Technology North Bangkok, Bangkok, Thailand
saranlita.c@email.kmutnb.ac.th

[2] Department of Information Technology Management, King Mongkut's University of Technology North Bangkok, Bangkok, Thailand
pym@kmutnb.ac.th

Abstract. Automatic questions and answers generated are challenging tasks that attract broad engagement. This paper presents an automatic question and answer generation from Thai sentences. We adopted question classification to predict the expected wh-question type before generating question and answer. This experimental study tested the question and answer generation on the sampling set of 300 Thai sentences based on the wh-question type that trained classify category from 15,000 samples. Automatic evaluation is performed on a reference set using the BLEU metrics and human evaluation. The experiment results show that the proposed method can generate questions and answers from Thai sentences. However, the performance based on the efficiency of Natural Language Processing includes tokenization, Named Entities Recognition, and Part of Speech tagging task because it could effectively extract focusing words correctly and suitable to answer the question that enhances the trustworthiness of future work in question and answer generation from Thai texts.

Keywords: Question and answer generation · Automatic question generation · Natural language processing · Thai sentences analysis

1 Introduction

Reading comprehension requires the ability to read and the questioning that readers should practice skills for improving learning performance. At present, efforts to create the ability for computers to understand texts by using question generation techniques from considering texts or information. This function could benefit information retrieval, a question answering system, automatic questioning, and cognition evaluation.

The main point of question and answer generated that could start by identify WH-question type (How, What, When, Where, and Who) from sentence. Thus, Question Classification (QC) is an important role, which the aim of helping guide question type from sentence. Natural language processing task would as tools to extract focusing words before transforming words into interrogative sentences.

P. Meesad et al. (Eds.): IC2IT 2022, LNNS 453, pp. 163–172, 2022.
https://doi.org/10.1007/978-3-030-99948-3_16

Thai questions are asked quite differently from English questions. Word order of Thai "wh" questions is different from English word order with another place.

There are many studies about automatic question generation nevertheless, few studies about question and answer generation in Thai texts. Thus, the contributions of our works were studied an automatic question and answer generation approach to generate questions from Thai sentences. There are challenges on study automatic question and answer generation in Thai texts which is lack of dataset for training and experiment on question answering in Thai texts and limitation of performance analysis Thai texts. This paper proposes a framework for question and answer generation from Thai sentences. In this work, we combined the Natural language processing step before going to the classification task for predict wh-question type predicted and applied Natural language processing annotation for focused answer possible based on question type predicted. Then we evaluated our method to analyze the acceptance of answers generated for their question.

The rest of this paper is organized as follows. In Sect. 2, we start with the literature reviews and related works. We show the implementation of our approach in Sect. 3. Section 3 shows the details of experimental results. Finally, Sect. 4 concludes the experiments and future work.

2 Related Works

Question generation basically started in 2008 with the National Science Foundation sponsored workshop on The Question Generation Shared Task and Evaluation Challenge [1]. Question generation (QG) is the task of automatically generating questions from inputs such as text, raw data, and knowledge bases. Further, the authors [2] defined question generation (QG) as the task of generating reasonable questions from an input, which can be structured or unstructured data. Question generation approaches can be classified into three categories 1) syntax-based, 2) semantic-based, and 3) template-based. Kurdi et al. [3], author introduced the approaches for question generation to two groups the level of understanding of the input required by the generation approach and the procedure for transforming the input into questions.

The primary step in generating automated questions is the data preprocessing stage, which requires Natural Language Processing (NLP) to understand the meaning. Zhou et al. [4], authors used POS and NER to consider choosing sentences, or words that will be useful to questions. Most recent research on automated question generation in Thai texts focuses on data preprocessing. Kwankajornkiet et al. [5] proposed a method for generating fill-in-the-blank questions with multiple choices from Thai text. Their results found that most unacceptable questions are hard to answer due to mistakes in extracting clauses from the text. Phakmongkol et al. [6] proposed a method for generating questions for Thai with Multilingual Text-to-Text Transfer Transformer.

2.1 Data

In this study, we used a Thai Wikipedia QA Corpus [7]. The corpus contains the factoid 15,000 question-answer pairs. We applied an above datasets for training question classification. Therefore, we first transformed interrogative sentence to simple sentence. Tokenization is a significant phase of Thai texts pre-processing. We applied the tokenization tools of PyThaiNLP [8], using dictionary-based word segmentation combined with the Newmm model and Wangchanberta for subword tokenize. The question type reflected by the interrogative word is that some question words have different words, but they have the same meaning shown in Table 1. The statistics of words in the dataset grouped by wh-question type are shown in Table 2.

Table 1. Number of words in the experimental dataset.

Question Type	Question words
"How" (เท่าไหร่ /tâo rài/)	"กี่, ที่เท่าไหร่, เท่าใด"
"What" (อะไร /à-rai/	"ใด, อะไร"
"When" (เมื่อไหร่ /m^eua r'ai/)	"เมื่อไหร่, เมื่อใด, เมื่อไร, วันที่เท่าใด, ปีค.ศ.ใด, ปีพ.ศ.ใด"
"Where" (ที่ไหน /têe năi/)	"ที่ไหน, ที่ใด, จังหวัด อะไร, ประเทศอะไร"
"Who" (ใคร /têe năi/)	"ใคร, ผู้ใด, คนใด"

Table 2. Number of words in the experimental dataset.

Wh question	Number of word	Average per sentence	Question classification
How	27,102	15.2429	1,778
What	111,209	14.3662	7,741
When	22,737	14.318	1,588
Where	22,634	12.7947	1,769
Who	28,976	13.6421	2,124
Overall	**212,658**	**14.0727**	**15,000**

2.2 Question Classification

Adding POS tagging on input to predict Wh-question class followed in the previous studies [9], POS tagging could be a benefit for resolving confusion sense for predict question class. For the data pre-processing step, input sentences were added more token sequence with POS by using the NLP toolkit for Thai (PyThaiNLP) [8] as in Algorithm 1. Formally, given a sequence of tokens in simple sentence (s) = $\langle w_1, w_2, ..., w_N \rangle$, we combine POS tagging to output a list of tuples, which is a POS tag mentioned in each tokens of sentence. $\langle p_1, p_2, ..., p_N \rangle$.

Algorithm 1: adding POS tags

Result: adding POS tag on simple sentence
initialization;
Read Sentence S_i from dataset; S is the set of sentences in the dataset.
while *each* S_i *in* S **do**
  Tokenize S_i and extract the set of term W_i; P_w is a POS tag of each the W_i in S_i;W_i is a set that contains all the word in S_i
  input = $S_i + P_{i(P_{ws},..,P_{we})}$;
  P_{ws} is the start POS tag and P_{we} is the end POS tag of W_i in S_i;
end

In this work, we split 80% as the training set (12,000 samples) and 20% (3,000 samples) as the testing set. We used Support Vector Machine (SVM) combined with Unigram+Bigram and TF-IDF. The accuracy of question classification when applied SVM model on adding POS input with the micro F_1-score of 0.8247. While the accuracy when applied SVM model on simple sentence input with the micro F_1-score of 0.8127. The model got an accuracy for predict the question type from Thai sentence that yielded the best performance with an F_1-score of 0.87 for class type "What" following by "When," "How," "Who," and "Where" with an F_1-score of 0.84,0.82,0.76 and 0.65, respectively. Then we exported a model that trained the model to make predictions wh-question type to generate questions and answers in the following step. The results showed that SVM (Unigram+Bigram with TF-IDF), on adding POS tag input better than the simple sentences. Then we applied SVM model that training on adding POS tags to predict the question type for question generate task.

2.3 Question and Answer Generation

In this step, question and answer generated by considering words that applied Named Entities Recognition (NER) [10] and Part-of-Speech (POS) tag [11] for focusing words that were possible answers from a sentence based on a question type. We conducted the condition for extract the focusing words with five rules.

Rule 1 Who category: search the NER tag as PERSON tag;
Rule 2 Where category: search the NER tag as LOCATION tag;
Rule 3 When category: search the NER tag as DATE or TIME tag;
Rule 4 How category: search the NER tag as LEN, MONEY or search the POS tag as NU tag;
Rule 5 What category: search the subject and object of the sentence;

Then we replaced the focusing words with question words. Figure 1 shows our purposed frameworks for question and answer generation. Table 3, 4, 5, 6 and 7 show that question and answer output when input defines question type.

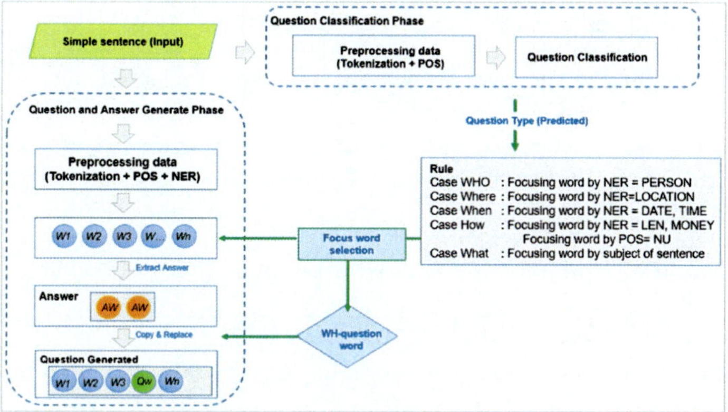

Fig. 1. An our purposed frameworks for question and answer generation

Table 3. Examples of generated question and answer (How)

Sentence: เกมบันไดงูต้องมีผู้เล่นตั้งแต่สองคนขึ้นไป /gem ban-dai ngoo dtông mee pôo lên dtâng dtàe sŏng kon kêun bpai/ (Snake and Ladder game requires two or more players.)	
Ground truth question:	เกมบันไดงูต้องมีผู้เล่นตั้งแต่กี่คนขึ้นไป /gem ban-dai ngoo dtông mee pôo lên dtâng dtàe gèe kon kêun bpai/ (Snake and Ladder game requires at least how many players?)
Ground truth answer:	สอง /sŏng/ (Two)
Predicted question:	เกมบันไดงูต้องมีผู้เล่นตั้งแต่เท่าไหร่คนขึ้นไป /gem ban-dai ngoo dtông mee pôo lên dtâng dtàe tâo
	rài kon kêun bpai/
Predicted answer:	[สอง][คน] /[sŏng][kon]/ (Two people)

Table 4. Examples of generated question and answer (Who)

Sentence: ผู้ประดิษฐ์ลูกโป่งยางได้เป็นคนแรกของโลกคือไมเคิลฟาราเดย์ /pôo bprà-dìt lôok bpòhng yaang dâai bpen kon râek kŏng lôhk keu/ mai-kern-faa-raa-day/ (The inventor of the world's first rubber balloon was Michael Faraday.)	
Ground truth question:	ผู้ประดิษฐ์ลูกโป่งยางได้เป็นคนแรกของโลกคือใคร /pôo bprà-dìt lôok bpòhng yaang dâai bpen kon râek kŏng lôhk keu krai/ (Who is the world's first rubber balloon invented?)
Ground truth answer:	ไมเคิลฟาราเดย์ /mai-kern-faa-raa-day/ (Michael Faraday)
Predicted question:	ผู้ประดิษฐ์ลูกโป่งยางได้เป็นคนแรกของโลกคือใคร /pôo bprà-dìt lôok bpòhng yaang dâai bpen kon râek kŏng lôhk keu krai/ (Who is the world's first rubber balloon invented?)
Predicted answer:	ไมเคิลฟาราเดย์ /mai-kern-faa-raa-day/ (Michael Faraday)

2.4 Evaluate Question and Answer Generate

The authors evaluated the test set that is randomly selected 300 samples (60 samples for each wh-question type) from the test set of 3,000 samples. Then we compared questions generated and ground truth question in the dataset with Bilingual Evaluation Understudy (BLEU). It explores for actual match of words

Table 5. Examples of generated question and answer (When)

Sentence: รถไฟฟ้าปารีสสาย 3 เปิดให้บริการในปีพ.ศ. 2447 /rót fai fáa bpaa-rêet săai săam bpèrt hâi bor-rí-gaan nai bpèep por · sŏr sŏng-pan-sèe-rói-sèe-sìp-jèt/ (Paris Metro Line 3 opened for service in B.E. 2447.)	
Ground truth question:	รถไฟฟ้าปารีส สาย 3 เปิดให้บริการในปี พ.ศ. อะไร /rót fai fáa bpaa-rêet săai săam bpèrt hâi bor-rí-gaan nai bpee por·sŏr à-rai/ (When is Paris Metro line 3 open for begin service?)
Ground truth answer:	2447 /sŏng-pan-sèe-rói-sèe-sìp-jèt/ (2447)
Predicted question:	รถไฟฟ้าปารีสสาย 3 เปิดให้บริการในปีเมื่อไหร่ /rót fai fáa bpaa-rêet săai săam bpèrt hâi bor-rí-gaan nai bpee mêuua rài/ (When is Paris Metro line 3 open for begin service?)
Predicted answer:	พ.ศ. 2447/por · sŏr sŏng-pan-sèe-rói-sèe-sìp-jèt/ (B.E.2447)

Table 6. Examples of generated question and answer (What)

Sentence: เริ่มแรกมหาวิทยาลัยกรุงเทพได้รับการก่อตั้งในชื่อว่าวิทยาลัยไทยเทคนิค /rêrm râek má-hăa wít-tá-yaa-lai grung têp dâai ráp gaan gòr dtâng nai /chêu wâa wít-tá-yaa-lai tai ték-ník/ (Initially, Bangkok University was established as the Thai Technical College.)	
Ground truth question:	เริ่มแรกมหาวิทยาลัยกรุงเทพได้รับการก่อตั้งในชื่อว่าอะไร /rêrm râek má-hăa wít-tá-yaa-lai grung têp dâai ráp gaan gòr dtâng nai chêu wâa à-rai/ (What was the name of Bangkok University originally established?)
Ground truth answer:	วิทยาลัยไทยเทคนิค/wít-tá-yaa-lai tai ték-ník/ (Thai Technical College)
Predicted question:	เริ่มแรกอะไรได้รับการก่อตั้งในชื่อว่าวิทยาลัยไทยเทคนิค /rêrm râek à-rai dâai ráp gaan gòr dtâng nai chêu wâa wít-tá-yaa-lai tai ték-ník/ (What was initially established as the Thai Technical College?)
Predicted answer:	มหาวิทยาลัยกรุงเทพ /má-hăa wít-tá-yaa-lai grung têp/ (Bangkok University)

Table 7. Examples of generated question and answer (Where)

<table>
<tr><td colspan="2">Sentence: อัฟกานิสถานมีอาณาเขตทางทิศตะวันตกจรดกับประเทศอิหร่าน /àf-gaa-nít-tăan mee aa-naa kèt taang tít dtà-wan dtòk jà-ròt gàp bprà-têt ì-ràan/ (Afghanistan is bordered to the west with Iran.)</td></tr>
<tr><td>Ground truth question:</td><td>อัฟกานิสถานมีอาณาเขตทางทิศตะวันตกจรดกับประเทศอะไร /àf-gaa-nít-tăan mee aa-naa kèt taang tít dtà-wan dtòk jà-ròt gàp bprà-têt à-rai/ (Afghanistan has what country to the west?)</td></tr>
<tr><td>Ground truth answer:</td><td>อิหร่าน/ì-ràan/ (Iran)</td></tr>
<tr><td>Predicted question:</td><td>อัฟกานิสถานมีอาณาเขตทางทิศตะวันตกจรดกับที่ไหน /àf-gaa-nít-tăan mee aa-naa kèt taang tít dtà-wan dtòk jà-ròt gàp têe năi (Where is Afghanistan's territory to the west?)</td></tr>
<tr><td>Predicted answer:</td><td>ประเทศอิหร่าน /bprà-têt ì-ràan/ (Iran)</td></tr>
</table>

and matches against a set of standard references. It computes the average n-gram score of size 1 to 4 for precision.

The BLEU metric lacks automatic fluency evaluation when replaced with only a few words in the sentence or other words that can be unacceptable or acceptable but not the same with ground truth sentences. In order to reflect grammatical mistakes in the generated sentence, we conducted the human evaluation with fluency as one of the criteria. In human evaluation, we discussed the evaluation criterion of questions with a different score (5 = All critical facts are present, 4 = Most of the critical facts are present, 3 = Some critical facts are missing, 2 = Most of the critical facts are missing, and 1 = All critical facts are missing). Five Thai people who use their mother tongue or native language in the Thai language decided to rank each question and answer generated.

$$Percentage(\%) = \frac{Average\ Score}{Total\ Number\ of\ Survey\ responses} \times 100 \qquad (1)$$

We calculated the average score and the acceptable percentage. To calculate the F_1-score, we count the number of words found in the ground truth question and the prediction. We also performed a human evaluation to analyze the acceptance of answers generated for their question. The number of points from each answer was generated to reach a total number and calculated the average score. The maximum possible score is 1, which is an answer to the question, while the lowest score is 0, which is not an answer.

3 Experiment Results

In this experiment, we have presented the method for generating questions and answer automatically. We evaluated our proposed method with the BLEU metric and human evaluation. Table 8 shows an automatic evaluation overall with a BLEU-4 score is 0.7415 and F_1-score is 0.7733. "When" category was the highest score with the BLEU-4 of 0.8403 and F_1 score is 0.8704. The lowest score is the "What" with 0.5186 and F_1-score is 0.6185.

As evident from Fig. 2, the human evaluation score showed the performance of our proposed could generate questions and answer. Humans are satisfied with the question and answers on the test set. The output is that most critical information is present, and they can infer the answer to the question with an average of 87.98%. The question's score on the "How" category is the highest, an average of 94.03%. In contrast, the score of questions in the "What" category is the lowest, an average of 78.65%. We found that the good answer of the question with an average of 93.59%. The answer's score on the "When" category is the highest, an average of 98.31%.

Table 8. Experimental results on the test set with automatic evaluation metrics

	BLEU-1	BLEU-2	BLEU-3	BLEU-4	Precision	Recall	F1
How	0.8516	0.8145	0.7872	0.7517	0.8078	0.8037	0.8042
What	0.6470	0.5972	0.5626	0.5186	0.6567	0.6012	0.6185
When	0.8874	0.8626	0.8508	0.8403	0.9062	0.8394	0.8704
Where	0.7656	0.7444	0.7285	0.7086	0.7958	0.7566	0.7695
Who	0.8575	0.8179	0.7759	0.7535	0.7986	0.8145	0.8042
Overall	0.8018	0.7673	0.7410	0.7145	0.7930	0.7631	0.7733

The results showed that the evaluated average scores by the judges were so different from the automatic evaluation metric because these words might be as diverse. There are two main reasons for the different questions and unanswerable. First, when a human asks a question based on a passage, he will consider the answer as a target related to different words in the sentence. Thus, when comparing the same input sentence for generating questions and answers. The output can ask various questions based on target or focusing words. Even though we assigned the wh-question type before generating questions and used NER and POS as tools for select words, we could still ask the different questions with a different word on acceptable. Second, The NER tags of answers correlated with the question type. We found that in some cases, NER are not perfect. Therefore, we might not selectable a focus word from a sentence. Thus, it was be a fail factor into question generate and answer.

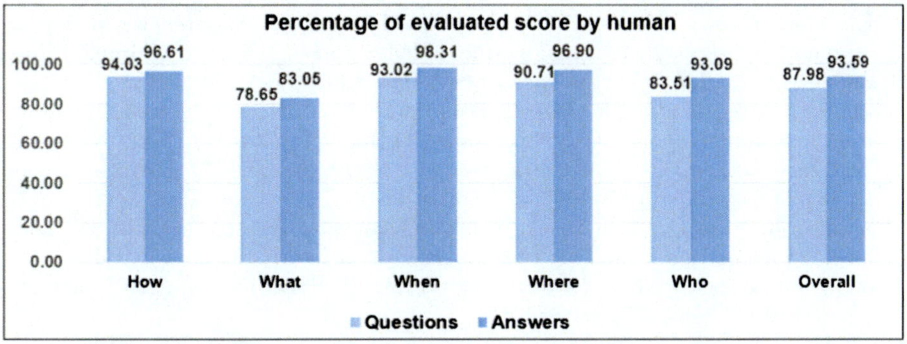

Fig. 2. Percentage of evaluated score by human

4 Conclusion and Future Work

In this paper, we presented the framework of automatic question and answer generation in Thai texts. In the first step, we built model for question classification applied SVM model with unigram+bigram and TF-IDF on adding POS in the input. Then, we investigated the rule for generate question and answer based on question type predicted. The proposed method is useful NLP annotation tags that are NER and POS.

The experiment results showed that our proposed led to good performance in question and answer generation for Thai sentences. Although our method requests a minor training set, deep learning requires many training data because of the enormous number of parameters needed to be tuned by a learning algorithm. The performance of Natural Language Processing affects the acceptable of questions and answers generated from Thai texts because they could effectively extract focusing words correctly and suitable to answer the question. Our future work will be focused on improve the performance of automatic question and answer from Thai texts to multi question possible. In addition, we are interested in investigate with deep learning and transformer models for question and answering generated from Thai texts.

Acknowledgements. We gratefully acknowledges support from the Office of the Permanent Secretary (OPS), Ministry of Higher Education, Science, Research and Innovation, Thailand. We thank King Mongkut's University of Technology North Bangkok in Thailand for support Information Technology.

References

1. Rus, V., Graesser, A.C.: Workshop Report: The Question Generation Task and Evaluation Challenge. Institute for Intelligent Systems, Memphis, TN (2009)
2. Yao, X., Bouma, G., Zhang, Y.: Semantics-based question generation and implementation. Dialogue Discourse **3**(2), 11–42 (2012)

3. Kurdi, G., Leo, J., Parsia, B., Sattler, U., Al-Emari, S.: A systematic review of automatic question generation for educational purposes. Int. J. Artif. Intell. Educ. **30**(1), 121–204 (2019). https://doi.org/10.1007/s40593-019-00186-y
4. Zhou, Q., Yang, N., Wei, F., Tan, C., Bao, H., Zhou, M.: Neural question generation from text: a preliminary study. In: Huang, X., Jiang, J., Zhao, D., Feng, Y., Hong, Yu. (eds.) NLPCC 2017. LNCS (LNAI), vol. 10619, pp. 662–671. Springer, Cham (2018). https://doi.org/10.1007/978-3-319-73618-1_56
5. Kwankajornkiet, C., Suchato, A., Punyabukkana, P.: Automatic multiple-choice question generation from Thai text. In: 2016 13th International Joint Conference on Computer Science and Software Engineering (JCSSE), pp. 1–6 (2016). https://doi.org/10.1109/JCSSE.2016.7748891
6. Phakmongkol, P., Vateekul, P.: Enhance text-to-text transfer transformer with generated questions for Thai question answering. Appl. Sci. **11**(21), 10267, 1–17 (2021). https://doi.org/10.3390/app112110267
7. Trakultaweekoon, K., Thaiprayoon, S., Palingoon P., Rugchatjaroen A.: The first Wikipedia questions and factoid answers corpus in the Thai language. In: 2019 14th International Joint Symposium on Artificial Intelligence and Natural Language Processing (iSAI-NLP), pp. 1–4 (2019). https://doi.org/10.1109/iSAI-NLP48611.2019.9045143
8. Phatthiyaphaibun, W., Chaovavanich, K., Polpanumas, C., Suriyawongkul, A., Lowphansirikul, L., Chormai, P.: PyThaiNLP: Thai Natural Language Processing in Python. Zenodo. https://doi.org/10.5281/zenodo.3519354
9. Chotirat, S., Meesad, P.: Part-of-speech tagging enhancement to natural language processing for Thai wh-question classification with deep learning. Heliyon **7**(10), 1–13 (2021). https://doi.org/10.1016/j.heliyon.2021.e08216
10. Phatthiyaphaibun, W.: wannaphongcom/thai-ner 1.2.1 (1.2.1). Zenodo (2019). https://doi.org/10.5281/zenodo.3342459
11. Boonkwan, P., et al.: The Annotation Guideline of LST20 Corpus. arXiv:2008.05055 (2020)

SPP-EEGNET: An Input-Agnostic Self-supervised EEG Representation Model for Inter-dataset Transfer Learning

Xiaomin Li and Vangelis Metsis(✉)

Texas State University, San Marcos, TX 78666, USA
{x_l30,vmetsis}@txstate.edu

Abstract. There is currently a scarcity of labeled Electroencephalography (EEG) recordings, and different datasets usually have incompatible setups (e.g., various sampling rates, number of channels, event lengths, etc.). These issues hinder machine learning practitioners from training general-purpose EEG models that can be reused on specific EEG classification tasks through transfer learning. We present a deep convolutional neural network architecture with a spatial pyramid pooling layer that is able to take in EEG signals of varying dimensionality and extract their features to fixed-size vectors. The model is trained with a contrastive self-supervised learning task on a large unlabelled dataset. We introduce a set of EEG signal augmentation techniques to generate large amounts of sample pairs to train the feature extractor. We then transfer the trained feature extractor to new downstream tasks. The experimental results[1] show that the first few convolutional layers of the feature extractor learn the general features of the EEG data, which can significantly improve the classification performance on new datasets (The source code on Github: https://github.com/imics-lab/eeg-transfer-learning).

Keywords: EEG · Self-supervised learning · Contrastive learning · Transfer learning · Spatial pyramid pooling

1 Introduction

Electroencephalography (EEG) has enabled numerous applications both in and out of the clinical domain. For example, it has been used for diagnosing epilepsy [1], monitoring sleep stages [2], emotion recognition [3], human-computer interaction [4], etc. However, EEG signals often have a low signal-to-noise ratio, high variability among subjects, and are collected in configurations that are incompatible across datasets, making them hard to interpret and challenging to use in an end-to-end automated processing pipeline (i.e., applying machine learning models on raw signals).

In the past, most machine learning EEG applications relied on hand-crafted features extracted by neuroscience experts. They usually required designing specific solutions for one particular task and are hard to be re-used on other

P. Meesad et al. (Eds.): IC2IT 2022, LNNS 453, pp. 173–182, 2022.
https://doi.org/10.1007/978-3-030-99948-3_17

tasks. Due to the difficulty of manually extracting useful features, only a limited amount of tasks with clear EEG patterns can benefit from automatically trained machine learning models. Deep learning allows end-to-end signal processing, feature extraction, and model training, which has enabled significant performance improvements in the EEG field [5–7] compared to traditional machine learning methods. Despite the success of deep learning in classifying EEG signals, the majority of these approaches are subject to the limited amounts of labeled EEG recordings that often lead to relatively shallow deep neural networks, trained in supervised ways, and on a single dataset.

Applications of deep learning in other domains (e.g. Computer Vision and NLP), have vastly benefited from transfer learning, i.e. starting with a neural network that has been pre-trained on big datasets and then fine-tuned on a specific task to achieve better classification accuracy rates and training performance speedup on smaller datasets.

Self-supervised Learning (SSL) is an approach for learning useful representations from unlabelled data. SSL allows deep neural networks to first learn general-purpose features from large unlabeled datasets and then fine-tune the representations for the supervised downstream tasks. Contrastive learning has been used to learn the general features of a dataset without labels by teaching the model which data points are similar or different. It has achieved great success in many fields. For example, image patches reordering [8] in Computer Vision, sentence embedding [9] in Natural Language Processing, and Wave2Vec [10] in time series data.

A few works have also implemented contrastive learning on EEG data. Banville et al. [11] developed three solutions mainly by defining negative windows and positive windows located in different places in an EEG recording and sample positive and negative data pairs from those windows. However, it only implements the downstream tasks on the same dataset trained for the pre-training task but did not verify if the learning features also fit other datasets. Mohsenvand et al. [12] developed a single-channel EEG representation model which fused multiple datasets and recombined multiple channels in one to boost the variety of channels. The work of Kostas et al. [13] is the most similar work compared to ours. They designed a CNN feature extractor and pass features to a transformer model to conduct contrastive learning tasks. However, the datasets used for the pre-training task and downstream tasks all need to be resampled to the same sampling rate.

In this paper, we present a new framework that allows us to, 1) learn EEG representations in a self-supervised manner via contrastive learning without requiring external labels, and 2) transfer the pre-trained feature extractor to other downstream tasks without modifying target dataset dimensions to match the properties of the original datasets used for the self-supervised learning task.

We design a contrastive learning task to train the feature extractor. To generate contrastive learning data pairs, we introduce a few EEG data augmentation methods. For example, adding noise, downsampling, random cropping, pooling, and quantizing signal voltage values, etc. To transfer the trained feature extractor without modifying the downstream dataset's input dimensions, we design a densely connected convolutional neural network with spatial pyramid pooling layer [14] to map any size of input data to fixed size vectors. Two data samples in a data pair both pass through the feature extractor and the absolute difference value are calculated in the latent space. Then a linear classifier is used to distinguish whether it is a positive data pair or negative one.

While transferring the pre-trained feature extractor to downstream tasks, we load the parameters of different numbers of convolutional layers to target classification models. The experimental results show that the contrastive learning task helps the feature extractor learn general EEG signal features that can be re-used on downstream tasks to improve the model's classification performance.

Our contributions can be summarized as follows:

- We exploit an EEG signal feature extractor that is agnostic of the input data dimension and can be re-used for any downstream task without modifying the target dataset's properties.
- We conduct a contrastive self-supervised learning task to extract general features from the EEG signals from massive EEG data without external labels.
- We perform transfer learning from the pre-trained feature extractor to other smaller EEG datasets. The experimental results show significant classification performance improvements compared to training smaller datasets on deep learning models from scratch.

The rest of the paper is organized as follows. Section 2 introduces our methodology. The datasets, data preprocessing steps, experiments setup, and results are described in Sect. 3. Section 4 concludes the paper.

2 Method

2.1 SPP-EEG Feature Extractor

We design an input dimension agnostic feature extractor, SPP-EEG network. The architecture is shown in Fig. 1. The convolution layers all have the kernel size $(1, 3)$ and filter count $F = 32$. They are applied to all input signal channels to learn the representations of each specific 10-20 EEG montage system channel. There are two convolution layers which have stride size $(1, 2)$ used to down-sample the feature map by a factor of $1/2$. A dropout layer is added after every three convolution layers to alleviate the over-fitting problem.

The Spatial Pyramid Pooling layer ensures every input signal length and shape be extracted to the exact same length of feature vectors. By choosing a pooling bin $[A, B, C]$, a vector of $size = F * (A^2 + B^2 + C^2)$ will be extracted after the spatial pyramid pooling layer for all shapes of input data. Then a dense layer with the embedding size 100 is added in this model. The spatial pyramid pooling layer enables the feature extractor to be trained with multiple EEG datasets with various data dimensions and learns the general EEG data features that co-exist on all recordings. And it can be transferred to any other downstream EEG classification task without modifying the original signal collection settings (e.g. sampling rate, event window size, etc.).

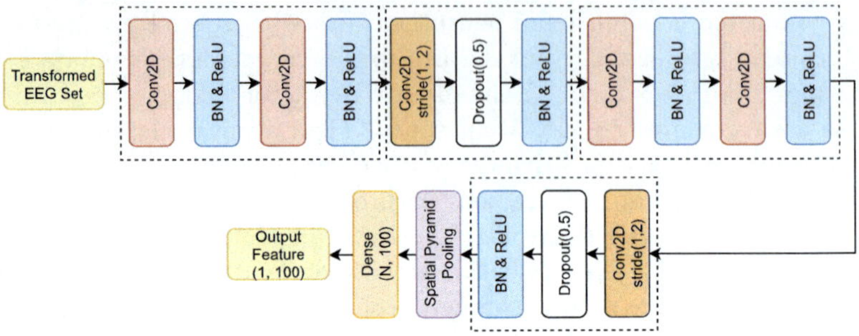

Fig. 1. SPP-EEG feature extractor architecture. Any dimensional input signals will be mapped to the same size 1D vectors. The Conv2D layers reserve general EEG signal features that can be transferred to downstream tasks.

2.2 Contrastive Learning

Figure 2 shows the contrastive learning pipeline. We denote the input multivariate time-series $S \in R^{1 \times C \times N}$, 1 represents that we view each EEG input as an 2D image with only one color channel. C is the total number of EEG data channels following the 10-20 EEG Montage system [15]. N is the total time-points of each input data sample. Here N can vary depending on various event window sizes, data transformations, and signal sampling rates, etc.

To produce labeled samples from the multivariate time-series S, we propose to sample pairs of time windows $(x_t, x_{t'})$ where each window x_t, $x_{t'}$ is in $R^{1 \times C \times N}$. The first window x_t is referred to as the "anchor window". The second window $x_{t'}$ can be a transformation of the anchor window, where $(x_t, x_{t'})$ has the label '1'. $x_{t'}$ can also be a transformation of a random time window which is far away from the anchor window for a distance of at least 10 times the window size, where $(x_t, x_{t'})$ has the label '0'. Our assumption is that an anchor window and its transformed counterparts still have some similar features and should have the same label, whereas the distant window should have a different label.

In order to learn how to discriminate pairs of time windows based on their sampled position, we use the feature extractor mentioned in Sect. 2.1 to map a window x to its representation in the latent space, $f_\Theta : R^{1 \times C \times N} \to R^D$. Ultimately we expect f_Θ to learn an informative representation of the raw EEG input which can be reused in different downstream tasks. A contrastive module g_{aug} is then used to aggregate the feature representations of each window. For our task, $g_{aug} : R^D \times R^D \to R^D$ combines representation form pairs of windows by computing an element-wise absolute difference, $g_{aug}(f_\Theta(x) - f_\Theta(x')) = |f_\Theta(x) - f_\Theta(x')| \in R^D$.

Then we use the binary cross entropy loss with a Sigmoid function on the predictions of g_{aug}, we can write a joint loss function as:

$$Loss = -\frac{1}{N}\sum_{i=1}^{N} y_i log\sigma(g_{aug}) + (1 - y_i) log(1 - \sigma(g_{aug}))$$

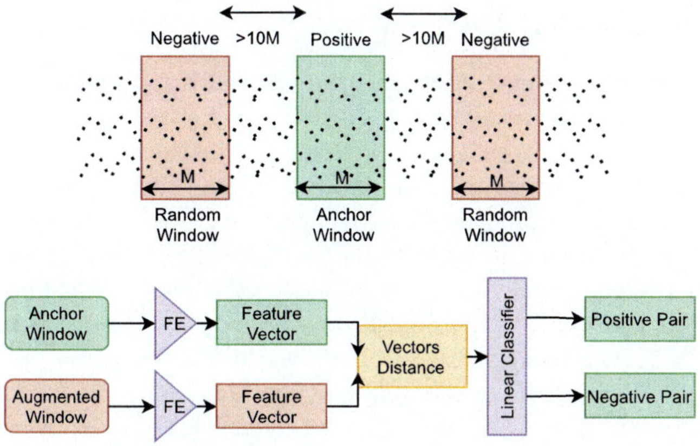

Fig. 2. Contrastive learning pipeline. FE represents the SPP-EEG feature extractor shown in Fig. 1. The absolute differences of two feature vectors are computed and a linear classifier is used to classify whether two vectors are similar or not in the latent space.

Table 1. Transformation ranges

Transformation	Min	Max
Add noise (scale)	0.02	0.1
Downsample (Hz)	100	256
Random crop (Time Points)	600	1200
Pooling (level)	2	5
Quantize (level)	10	30

2.3 Augmentation Methods

Data augmentation serves a dual purpose in our methodology. It allows us to generate more training data and it enables the self-supervised contrastive learning process, as explained in Sect. 2.2. Five different signal augmentation methods are used. Table 1 shows the range of each transformation. **Add Noise:** Add independent and identically distributed random noise to the input signal with the standard deviation of the random noise equal to a random generated scale. **Downsample:** Downsample input signals to a random sampling rate within a range but not violate the Nyquist-Shannon sampling theorem [16]. **Random Crop:** Select random sub-sequences in a range from the input signal. **Pooling:** Reduce the temporal resolution to original input 1/level without changing the input signal length. **Quantize:** Round each timepoint value to the nearest level in a level set.

2.4 Transfer Learning

After training the feature extractor with the contrastive self-supervised learning task, we transfer the feature extractor to different downstream EEG classification tasks and fine-tune it on the new dataset. The downstream classification model has a similar architecture with the pre-trained feature extractor, except we switch the spatial pyramid pooling layer to a maxpooling layer. The pooling size is decided upon each dataset's input window size. Such a decision is made empirically as the experimental results show the downstream tasks cannot benefit from the spatial pyramid pooling layer and it is also unnecessary to use it for single dataset classification tasks. Then we add a few dense layers and an output layer. The output dimension depends on each downstream task.

3 Experiments

3.1 Datasets

We conduct a self-supervised contrastive learning task on the TUH Normal/Abnormal dataset [17]. This dataset contains 2,993 recordings of around 15 min from 2,329 different patients who underwent a clinical EEG in a hospital setting. The training set contains 1,371 recordings labeled as normal and 1,346 labeled as abnormal. The evaluation set contains 150 and 126 recordings respectively.

We conduct transfer learning to three downstream datasets. The first one is from the same dataset used for feature extractor training. We select 30 normal and 30 abnormal patients' recordings that are not been used formerly. The second one is the EEG Motor Movement/Imagery Dataset [18,19], which consists of over 1500 one- and two-min EEG recordings, obtained from 109 volunteers. The recordings are labeled as $0, 1$ for four different binary motor imagery tasks. We used the first 50 subjects recordings in transfer learning experiments. The third one is the subset the first one, chooses 10 EEG recordings from the TUH normal/abnormal dataset in each training round to make up the training set and testing set.

3.2 Data Preprocessing

For the TUH Normal/Abnormal dataset, we use the same train/test split, to train and test the feature extraction model, as in the original paper [17]. For the transfer learning datasets, we use KFold $= 5$ cross-validation.

To preprocess the TUH dataset, we follow the EEG montage 10-20 system to select and reorder all recording channels to [FP1, FP2, F7, F3, FZ, F4, F8, T3, C3, CZ, C4, T4, T5, P3, PZ, P4, T6, O1, O2]. We discard the first 1-min time points from all recordings and crop the total length of a recording to no more than 20 min. We apply a (0.3-80 Hz) fifth-order band-pass Butterworth filter [20] to all recordings. 6-s time windows are extracted and each time window has the dimension $size = (1, 19, 6 * Hz)$. Then all time windows are channel-wise normalized.

For the transfer learning datasets, we select and reorder recording channels the same as the TUH dataset. We follow the original event time and the sampling rate of each dataset to divide time windows, which yields different sizes of time windows for different datasets. To all recordings we apply the filter bands suggested in their manuals. And similarly, all time windows are normalized in each channel.

3.3 Training Details

We conduct all experiments on an Intel server with 377 GB CPU memory and 2 Nvidia 1080 GPUs. The feature extraction model is trained for 100 epochs and saved at the best performance checkpoint. We use Adam optimizer [21] with the initial learning rate of 0.001 and batch size of 32. The learning rate degrades to its 95% after every 10 epochs. Since the contrastive data pairs we generated have different window sizes, we pad all time windows to the same size in a training batch to maximize the GPU's parallelization ability. For transfer learning tasks, we still use Adam optimizer with the learning rate 0.0001 or 0.00001 depending on the dataset. The training epochs are set to 50 in each fold and the batch size is set to 32 or 64 depending on the available GPU memory.

3.4 Transfer Learning Results

We first transfer the feature extractor to the TUH Normal/Abnormal dataset. As the first dataset described in Sect. 3.1. We segment the recordings to 6-second time windows and each is labeled as normal or abnormal. Then reload the parameters of different numbers of convolutional layers and train the classifier mentioned in Sect. 2.4. The experiment results are shown in Fig. 3. The red line is the classification accuracy obtained when training the same network architecture on the downstream dataset from scratch. The blue line shows the classification accuracy obtained on the downstream task when transferring the first n convolutional layers trained on the self-supervised learning task. From the plot, we can see the more than 6% accuracy improvement is achieved when transferring the first three layers on the TUH dataset. Overall, the accuracies obtained using our transfer learning approach are better than training from scratch except when transferring all six layers, which expected as in that case the original model is highly specialized on the initial self-supervised learning task. In other words, the last layers extract the patterns that particularly fit the former task which are not useful for downstream tasks.

Similar results can be found in Fig.4 when transferring the feature extractor to EEGBCI dataset which is a new dataset, not seen during the self-supervised pre-training process. The best performance appears when transferring the first four convolutional layers with around 4-5% accuracy improvement compared to training from scratch.

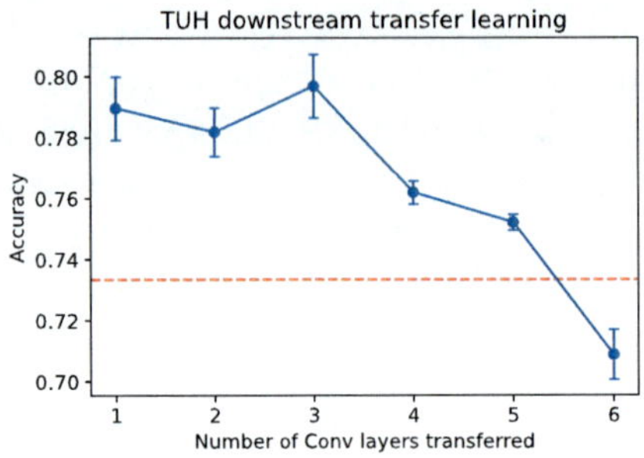

Fig. 3. TUH Downstream transfer learning results. The red line is the classification accuracy obtained when training the same network architecture on the downstream dataset from scratch. The blue line shows the classification accuracy obtained on the downstream task when transferring the first n pre-trained convolutional layers.

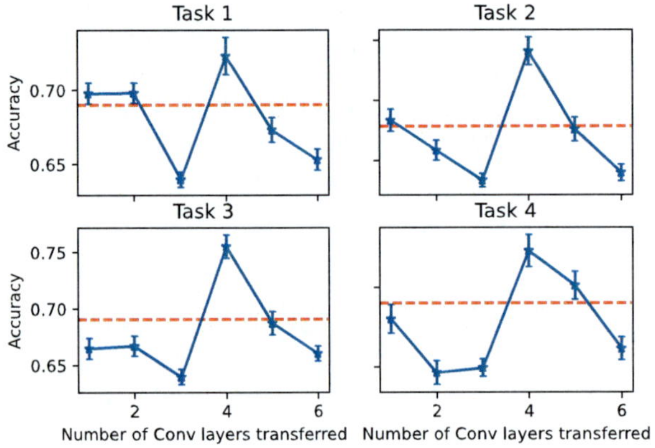

Fig. 4. EEGBCI transfer learning results for the four different tasks.

We conduct another experiment to show how the pre-trained feature extractor could benefit the downstream task on very small training sets. In this case, due to shortage of training data, the deep learning model quickly over-fits the training set and performs very poorly on testing set. The experiment uses only 10 EEG recordings from the TUH normal/abnormal dataset to make up the training set and testing set. We transfer the pre-trained feature extractor, training on three different training sets, and test on a single testing set. Figure 5 shows the transfer learning results. The smaller the difference between training and testing accuracy in the downstream deep learning model indicates less over-fitting, and the testing accuracy improves. From the plots, (1) we can observe that transferring the first three convolution layers parameters to the downstream task and freezing the first layer's parameters give the best practical result. (2) On such a small training set, the pre-trained feature extractor overall performs better than training from scratch. Our findings show that the number of layers to be transferred and the number of layers to be frozen when training on the downstream task matters. Transferring too many or freezing too many layers adds stricter restrictions on the downstream model which makes it performs worse on the testing set.

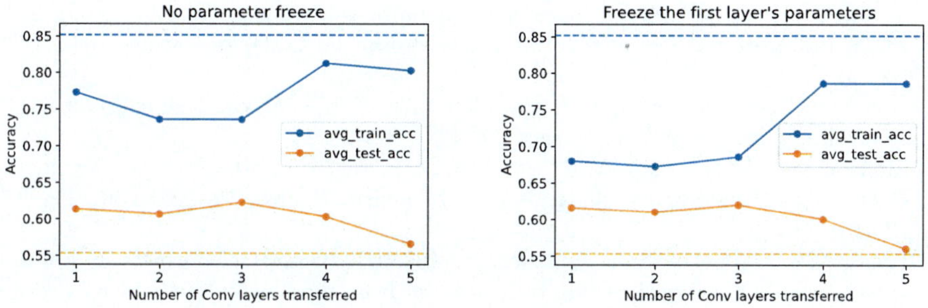

Fig. 5. Transfer learning results on a very small dataset. The upper and lower borderlines are the training from scratch training and testing accuracy.

4 Conclusion

In this work, we introduce an input data shape agnostic deep convolutional neural network for learning EEG data representations. We train the model with a self-supervised contrastive learning task and transfer the model to other smaller datasets with different recording setups. The experimental results verify the ability of contrastive learning to learn EEG data inner representation without external labels. And the pre-trained parameters can be used in other small EEG datasets as the initial training parameters to improve the downstream classification performance.

References

1. Acharya, R.U., et al.: Automated EEG analysis of epilepsy: a review. Knowl. Syst. **45**, 147–165 (2013)
2. Aboalayon, K.A.I., et al.: Sleep stage classification using EEG signal analysis: a comprehensive survey and new investigation. Entropy **18**(9), 272 (2016)
3. Nafjan, A.A., et al.: Review and classification of emotion recognition based on EEG brain-computer interface system research: a systematic review. Appl. Sci. **7**(12), 1239 (2017)
4. Lotte, F., Bougrain L., Clerc, M.: Electroencephalography (EEG)-based brain-computer interfaces (2015). https://doi.org/10.1002/047134608X.W8278
5. Schirrmeister, R.T., et al.: Deep learning with convolutional neural networks for EEG decoding and visualization. Hum. Brain Map. **38**(11), 5391–5420 (2017)
6. Lawhern, V.J., et al.: EEGNet: a compact convolutional neural network for EEG-based brain-computer interfaces. J. Neural Eng. **15**(5), 056013 (2018)

7. Wu, D., et al.: Transfer learning for motor imagery based brain-computer interfaces: a complete pipeline. arXiv preprintarXiv:2007.03746. (2020)
8. Falcon, W., Cho K.: A framework for contrastive selfsupervised learning and designing a new approach. arXiv preprintarXiv:2009.00104 (2020)
9. Gao, T., Yao, X., Chen D.: SimCSE: simple contrastive learning of sentence embeddings. arXiv preprintarXiv:2104.08821 (2021)
10. Yuan, Y., et al.: Wave2vec: learning deep representations for biosignals. In: 2017 IEEE International Conference on Data Mining (ICDM), pp. 1159–1164. IEEE (2017)
11. Banville H., et al.: Uncovering the structure of clinical EEG signals with self-supervised learning. J. Neural Eng. **18**(4), 1–32 (2021)
12. Mohsenvand, M.N., Izadi, M.R., Maes, P.: Contrastive representation learning for electroencephalogram classification. Mach. Learn. Health. PMLR. **136**, 238–253 (2020)
13. Kostas, D., Ouellette, S.A., Rudzicz, F.: BENDR: using transformers and a contrastive self-supervised learning task to learn from massive amounts of EEG data. arXiv preprint arXiv:2101.12037 (2021)
14. He, K., et al.: Spatial pyramid pooling in deep convolutional networks for visual recognition. IEEE Trans. Pattern Anal. Mach. Intell. **37**(9), 1904–1916 (2015)
15. Oostenveld, R., Praamstra, P.: The five percent electrode system for high-resolution EEG and ERP measurements. Clinic. Neurophys. **112**(4), 713–719 (2001)
16. Nemirovsky, J., Shimron E.: Utilizing bochners theorem for constrained evaluation of missing fourier data. arXiv preprintarXiv:1506.03300 (2015)
17. Lòpez S., Obeid I., Picone J.: Automated interpretation of abnormal adult electroencephalograms. PhD thesis. Temple University Graduate Board (2017)
18. Wolpaw, J.R., et al.: BCI2000: a general purpose brain-computer interface system. Soc. Neurosci. Abstr. **26**, 1229 (2000)
19. Goldberger, A., et al.: Components of a New Research Resource for Complex Physiologic Signals. PhysioToolkit, and Physionet, PhysioBank (2000)
20. Butterworth, S., et al.: On the theory of filter amplifiers. Wireless Eng. **7**(6), 536–541 (1930)
21. Kingma, D.P., Ba, J.: Adam: a method for stochastic optimization. arXiv preprint arXiv:1412.6980 (2014)

Enhancing Sentiment Analysis Using Hybrid Deep Learning

Watthana Ukaihongsar(✉) and Watchareewan Jitsakul

Faculty of Information Technology and Digital Innovation, King Mongkut's University of Technology North Bangkok, Bangkok, Thailand
s6307021858104@email.kmutnb.ac.th,
watchareewan.j@itd.kmutnb.ac.th

Abstract. The objective of this research is to enhance sentiment analysis of digital currency investors on twitter with hybrid deep learning. By using two deep learning algorithms, which are Convolutional Neural Network (CNN) and Gated Recurrent Unit (GRU) is called CNN-GRU. In this work, data from twitter textual content in English languages and Thai languages 1,000 samples were divided into positive message 500 samples and negative message 500 samples. Then prepare data before modeling as tokenization by Attacut algorithm, lower case transformation, stemmer and spilt data 70% to training data and 30% to testing data and also compared by adjust parameters to measure the classification efficiency. The experiment results showed that CNN-GRU was the best performance in the classification of positive message and negative message with Word Embedding Dimension 64, Number of Kernel 128 and Kernel Size 5, Memory 16, and recurrent dropout 0.4. The best result of CNN-GRU was accuracy 82.67%, precision 0.84, recall 0.80, f-measure 0.83, and ROC 0.82.

Keywords: Sentiment classification · Deep learning · Convolutional Neural Network · Gated Recurrent Unit

1 Introduction

Money is an essential factor in human life. It can be used as an intermediary in the exchange of goods, services, and can be used legally for liabilities. Currently, money is in the form of coins and banknotes but technology is evolving so rapidly that cause today has innovations are digital currencies. It doesn't have a physical form and is stored or exchanged on computer systems. There cannot be controlled by from government or banks. digital currencies have been increasingly popular. A lot of people are interested in future earnings speculation. Investors are attentive to invest when receiving positive news and positive comment on social media more than rather than consideration reliable sources [1]. Text mining with deep learning algorithms can adoption to sentiment classification that assessment may be used as an accessory factor for making investment decisions.

P. Meesad et al. (Eds.): IC2IT 2022, LNNS 453, pp. 183–193, 2022.
https://doi.org/10.1007/978-3-030-99948-3_18

Many deep learning algorithms have been applied in sentiment analysis such as Convolutional Neural Network classifier to analyze Chinese text sentiment [2], Chinese customer reviews products of consumer electric company using Gated Recurrent Unit [3], Latent Dirichlet Allocation (LDA) to train topic distribution of teat and used GRU-CNN classifier to classify web review, notebook computer review, and Chinese microblog sentiment [4] and Big Data analytic for tourism: A case study from Ubon Ratchathani Province, Thailand [5]. Each research is integrated deep learning algorithms for sentiment classification.

Therefore, this research has the purpose to apply deep learning algorithms to sentiment classification from textual content about digital currency on twitter by using three deep learning algorithms CNN, GRU, and CNN-GRU. Comparison performance with accuracy, precision, recall, f-measure, and ROC.

2 Literature Reviews

2.1 Convolutional Neural Networks (CNN)

Convolutional Neural Networks [6] are simulates human vision for feature extraction by using patterns in images to recognize objects. It at the area as a sub-area and bringing a group of sub-area to merge for extraction. CNN is suitable for images but can be used to classify non-image too.

2.2 Gated Recurrent Unit (GRU)

Gated Recurrent Unit [7] as a type of recurrent neural network. The GRU mechanism is able to process memories storing previous inputs in the internal state and can set how much to forget previous inputs for the next state. GRU is suitable for sequential data and fast processing by using less parameter.

2.3 Related Work

Many classification algorithms have been applied in sentiment analysis for the prediction of user emotion. Weng, L., Li, Q., and Xuehai, D. [8] developed a classification model for finding corresponding categories from movie review sentences, customer reviews about various products, and medical records. In this work, CNN, LSTM and GRU were used and propose a new algorithm IF-CGNN which combines CNN and GRU with initialized filters. The comparisons show that IF-CGNN gives the highest accuracy 72.5% Huang, C., and Liu, G. [9] propose to comparison traditional machine learning models LR and Support Vector Machine with deep learning model GCNN model base on GRU and CNN for sentiment analysis. Dataset include Chinese hotel comments (ChnSentiCorp) corpus and IMDB corpus. Comparative experimental results GCNN give the highest accuracy 94.9% on ChnSentiCorp corpus and highest accuracy 90% on IMDB corpus. Gaye, B., Zhang, D., and Wulamu, A. [10] did research on sentiment analysis of tweets from "Sentiment140". It consists of 1.6 million tweets among which 50% are positive and 50% are negative tweets. In this study, They are proposed LR-LSTM and compared performance with other models such as logistic regression, AdaBoost, random forest,

CNN, GRU, LSTM and CNN-LSTM. The LR-LSTM gives the highest classifier accuracy 99%, precision 99%, recall of 99%, and F1 Score 99%. Wang, X., Jiang, W. and Luo, Z. [11] studied on sentiment learning from three benchmarks are movie review (MR), Stanford Sentiment Treebank (SST1), SST2 same SST1 but remove neutral reviews and binary labels. Comparison with variants of the model, CNN-GRU-word2vec, CNN-LSTM-word2vec, CNN-GRU-rand, CNN-LSTM-rand, and AGV-GRU-word2vec. The results showed that CNN-GRU-word2vec was higher accuracy for MR, SST2 dataset and CNN-LSTM-word2vec was the best accuracy for SST1. In addition, Poomka, P. [12] developed a predictive model for sentiment analysis product review dataset from the site www.amazon.com. In this work, DNN-TF-IDF, CNN1D-TF-IDF, DNN-Doc2Vec, CNN1D- Doc2Vec, CNN2D-Word Embedding, LSTM- Word Embedding, and GRU-Word Embedding were used. Comparative results on reviews comments based on book, dvd, electronics, kitchen, music, and video. The DNN-TF-ID gives highest accuracy 0.84, precision 0.84, recall 0.83, and time to train 5.12 s. Moreover, Alotaibi, F. M., Asghar, M. Z., and Ahmad, S. [13] studies psychopathic classification to build an effective deep learning model for the detection of psychopaths. Dataset was collected from the social network using hashtag (#psychopath) and manually label the dataset by assigning a class label, i.e. "psychopath" or "non-psychopath," total 601 tweets, divide into 300 psychopaths and 301nonpsychopath. Comparison different classifiers models such as SVM, RF, NB, CNN, LSTM and BiLSTM. The comparisons show that CNN-LSTM gives the best results accuracy 91.67%, precision 0.93, recall 0.92, and F-measure 0.91. Furthermore, Zouzou, A., and El Azami, I. [14] propose to use GloVe as a word embedding, Adam optimizer, Adadelta optimizer and CNN, GRU and GRU-CNN model applied to movie review from IMDB dataset consist of 50k review. The experimental result that CNN-GRU-Adadelta optimizer model is best classification results with accuracy in training 86.34% and accuracy in testing 82.25%.

Based on previous studies, most of the classification process in sentiment analysis follows the single model but if a model is created that combine two or more model, the result will be better. In this research, we enhance classification accuracy by hybrid deep learning. By using two deep learning algorithms, which are Convolutional Neural Network (CNN) and Gated Recurrent Unit (GRU) is called CNN-GRU for sentiment analysis of digital currency investors on twitter.

3 Research Methodology

In this paper, the objective of this research is sentiment classification of digital currencies investors from twitter textual English and Thai language by using three deep learning algorithms. The research methodology can explain in Fig. 1.

From Fig. 1. The architecture of research methodology can explain into Sect. 3.1) Collect Data and Preprocessing and Sect. 3.2) Modeling and Evaluation Parameter.

3.1 Collect Data and Preprocessing

In this research, we collect data on opinions about digital currency on twitter in English text and Thai text from the website www.mediatoolkit.com and tweepy. By using keywords to collect data and hashtags include Bitcoin, Ethereum, Cardano, Binance coin, Solana, Dogecoin, BTC, ETH, ADA, BNB, SOL, and DOGE total 12,754 message.

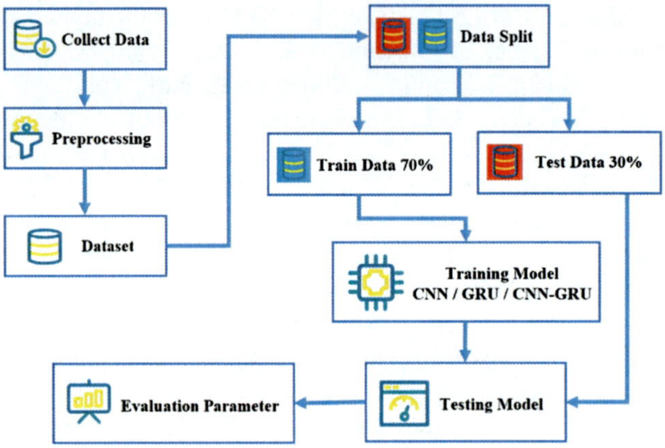

Fig. 1. The architecture of research methodology.

The researchers selected a total of 1,000 messages for testing, divided into 500 positive comments and 500 negative comments. Then prepare these data before modeling and manually label the dataset by assigning a class label for "Positive" and "Negative" as shown in Table 1.

Table 1. Labeled data example.

Texts	Sentiment
Nothing can stop #crypto Nothing can stop #bitcoin Be bullish	Positive
แต่ประเทศแถวนี้มอง #bitcoin เป็นฟองสบู่บ้างล่ะแชร์ลูกโซ่บ้างล่ะ นึกแล้วก็ขำดี	Negative
Money can't buy happiness. But Bitcoin can.	Positive
เค้าจะหลงรักอะไรก็หลงไป แต่ผมหลงรัก .....#btc	Positive
#Bitcoin fuck the bears. That's all.	Negative
Fuck idiot Sony sell #Bitcoin	Negative
I'm always #bullish on #cardano	Positive
Bad girls like #Bitcoin	Negative
#Bitcoin มักมีเซอร์ไพรส์เสมอ	Positive
Bears Waiting For the Market Crash #Bitcoin $BTC	Negative
#Bitcoin goddess must b so happy🤑 https://t.co/6sql8F4jLE	Positive
จะย่อตัวลงถึงขนาดไหนนั้นน #Bitcoin https://t.co/tEAZCdW9yk	Negative
That's easy, #ADA 🚀 🙂 https://t.co/qWpBhAKMEt	Positive
#ETH fees are so low, I'm not sure	Negative
#Ethereum New all time high ! 🚀🚀🚀🚀	Positive

From Table 1. The data are in English, Thai, symbols and special characters. Therefore, it is necessary to bring these data through the preparation process, tokenization, lower case transformation, removal stop words, removal special characters, stemming, and convert to numeric as "Positive" is 1, "Negative" is 0. Attacut Algorithm was used in this process because it can tokenize words effectively in both English and Thai languages as well as takes less time to process. The results as shown in Table 2.

Table 2. Example data after preprocessing.

Texts	Sentiment
['noth', 'can', 'stop', 'crypto', 'can', 'stop', 'bitcoin', 'bullish']	1
['รัฐ', 'แถว', 'bitcoin', 'ฟอง', 'สบู่', 'ล่ะ', 'แชร์', 'ลูก', 'โซ่', 'ล่ะ', 'นึก', 'ขำ', 'ดี']	0
['money', 'cant', 'buy', 'happi', 'but', 'bitcoin', 'can']	1
['เค้า', 'หลง', 'รัก', 'หลง', 'ผม', 'หลง', 'รัก', 'btc']	1
['bitcoin', 'bear', 'fuck']	0
['fuck', 'idiot', 'soni', 'sell', 'bitcoin']	0
['I', ''m', 'alway', 'bullish', 'cardano']	1
['bad', 'girl', 'like', 'bitcoin']	0
['bitcoin', 'เซอร์ไพรส์', 'เสมอ']	1
['for', 'market', 'crash', 'BC']	0
['bitcoin', 'goddess', 'must', 'b', 'happi']	1
['ย่อ', 'ตัว', 'ขนาด', 'bitcoin']	0
['hat'', 'easi', 'ada']	1
['EH', 'fee', 'low', 'Im', 'sure']	0
['ethereum', 'new', 'time']	1

Table 2. After data has been successfully prepared by preprocessing. Data in the form of tokenize, lower case, stemming and removal of space, symbols, and special characters. Sentiment labels are converted to numeric "1" and "0" in preparation for modeling.

3.2 Modeling and Evaluation Parameter

In this study, three deep learning models are applied to sentiment classification including CNN, GRU, and CNN-GRU. Each performance comparison by adjust parameters to find the best results with the same data set.

After preprocessed data then split divided into a training set and testing set with a ratio of 70:30. Training set amount 700 samples (70%) and Testing set amount 300 samples (30%). Setting considered parameters for three deep learning algorithms and adjust the parameters of each algorithm for performance comparison with accuracy, precision, recall, f-measure, and ROC. Considered parameters are shown in Table 3 and 4.

Table 3. Setting CNN considered parameter.

Considered parameter	Value
Word embedding dimension	64, 128, 256, 512
Number of Kernel	8, 16, 32, 64, 128, 256
Kernel size	3, 4, 5

Table 3. Setting considered a parameter for CNN model were adjusted to perfomance test as w_dim was 64, 128, 256, 512 Number of Kernel (n_kernel) was 8, 16, 32, 64, 128, 256 and Kernel Size (kernel_size) was 3, 4, 5.

Table 4. Setting GRU considered parameter.

Considered parameter	Value
Word embedding dimension	64, 128, 256, 512
Memory	8, 16, 32, 64, 128, 256, 512
Recurrent dropout	0, 0.1, 0.2, 0.3, 0.4, 0.5, 0.6, 0.7, 0.8, 0.9

Table 4. Setting considered parameter for GRU model were adjusted to performance test as w_dim was 64, 128, 256, 512 Memory was 8, 16, 32, 64, 128, 256, 512, and Recurrent Dropout (reccurrent_dropout) was 0, 0.1, 0.2, 0.3, 0.4, 0.5, 0.6, 0.7, 0.8, and 0.9.

Another parameter in the model is set to a constant. Example of another parameter is shown in Table 5.

Table 5. Example of another parameter.

Parameter	Value
Activation function Conv1D	relu
Kernel_regularizer L2	0.001
Dropout	0.25
First dense layer	64, sigmoid
Second dense layer	1, sigmoid
Adam optimizer rate	0.0005
Epochs	50
Batch_size	10

Table 4. Setting activation function Conv1D "relu", Kernel_regularizer L2 "0.001", Dropout "0.25", First Dense Layer "64 (Active Learning = sigmoid)", Second Dense

Layer "64 (Active Learning = sigmoid)", Adam Optimizer Rate "0.0005", Epochs "50", and Batch_size "10".

4 Experimental Results

In this section, evaluate the effective of three deep learning CNN, GRU, and CNN-GRU.

4.1 CNN

In this experiment, Considered parameters were adjusted to performance test as Word Embedding Dimension (w_dim) was 64, 128, 256, 512 Number of Kernel (n_kernel) was 8, 16, 32, 64, 128, 256 and Kernel Size (kernel_size) was 3, 4, 5. Evaluation Parameter with accuracy (Acc), precision (P), recall (R), f-measure (F) and ROC, The best result is shown in Table 6.

Table 6. The best result of CNN.

Parameter			Acc	P	R	F	ROC
w_dim	n_kernel	kernel_size					
64	128	5	80.33	0.80	0.81	0.81	0.80

Table 6 shows the best result of CNN with w_dim = "64", n_kernel = "128" and kernel_size = "5". The result was accuracy 80.33%, precision 0.80, recall 0.81, f-measure 0.81 and ROC 0.80.

4.2 GRU

In this experiment, considered parameters were adjusted to performance test as Word Embedding Dimension (w_dim) was 64, 128, 256, 512 Memory was 8, 16, 32, 64, 128, 256, 512, and Recurrent Dropout (reccurrent_dropout) was 0, 0.1, 0.2, 0.3, 0.4, 0.5, 0.6, 0.7, 0.8, and 0.9. Evaluation Parameter with accuracy (Acc), precision (P), recall (R), f-measure (F) and ROC, The best result is shown in Table 7.

Table 7. The best result of GRU.

Parameter			Acc	P	R	F	ROC
w_dim	memory	reccurrent_dropout					
128	16	0.4	82.33	0.81	0.83	0.82	0.82

Table 7 shows the best result of CNN with w_dim = "128", memory = "64" and reccurrent_dropout = "0.4". The result was accuracy 82.33%, precision 0.81, recall 0.83, f-measure 0.82, and ROC 0.82.

4.3 CNN-GRU

Experiment setting on a cascaded layer between CNN and GRU. Parameter was chosen from the best results from CNN and GRU. For the CNN layer select the number of Kernel and Kernel Size. GRU layer select Memory and Recurrent Dropout. The Considered parameter is Word Embedding Dimension (w_dim). Setting CNN-GRU Considered parameter and CNN-GRU diagram as shown in Fig. 2.

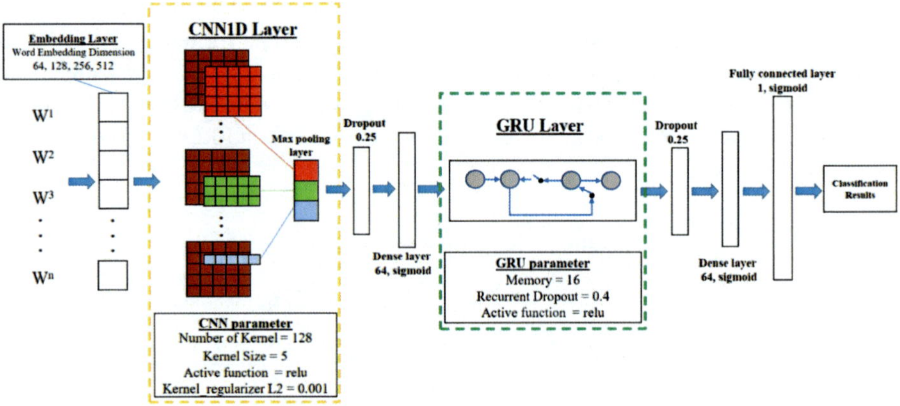

Fig. 2. CNN-GRU diagram

From Fig. 2, considered parameter was adjusted to performance test as Word Embedding Dimension (w_dim) was 64, 128, 256, 512 and other parameters are fixed, then repeat experiments 10 times. The best result is shown in Table 8.

Table 8. The best result of CNN-GRU.

Parameter	Acc	P	R	F	ROC
Word embedding dimension					
64	80.67	0.78	0.85	0.82	0.81
64	**83.00**	**0.85**	**0.80**	**0.82**	**0.83**
64	81.00	0.85	0.75	0.80	0.81
64	79.33	0.78	0.82	0.80	0.81
64	80.67	0.79	0.84	0.81	0.81
64	81.33	0.79	0.85	0.82	0.79
64	79.67	0.79	0.80	0.80	0.81
64	82.33	0.82	0.83	0.82	0.81
64	80.33	0.81	0.80	0.80	0.80

(continued)

Table 8. (*continued*)

Parameter	Acc	P	R	F	ROC
Word embedding dimension					
64	81.00	0.80	0.83	0.81	0.82
128	80.00	0.79	0.82	0.80	0.80
128	79.33	0.76	0.85	0.81	0.81
128	79.33	0.78	0.81	0.80	0.82
128	79.00	0.78	0.81	0.79	0.81
128	79.67	0.77	0.85	0.81	0.81
128	80.33	0.77	0.86	0.81	0.80
128	80.33	0.79	0.83	0.81	0.80
128	79.00	0.77	0.82	0.80	0.81
128	79.33	0.77	0.83	0.80	0.82
128	78.67	0.78	0.79	0.79	0.80
256	80.67	0.80	0.81	0.81	0.79
256	79.33	0.78	0.82	0.80	0.79
256	80.00	0.81	0.79	0.80	0.79
256	79.00	0.78	0.81	0.79	0.80
256	79.33	0.80	0.79	0.79	0.80
256	80.33	0.78	0.85	0.81	0.80
256	80.67	0.82	0.79	0.80	0.79
256	79.33	0.81	0.76	0.79	0.79
256	80.33	0.79	0.82	0.81	0.79
256	81.00	0.80	0.82	0.81	0.81
512	79.33	0.78	0.83	0.80	0.79
512	79.33	0.80	0.79	0.79	0.80
512	78.67	0.78	0.80	0.79	0.79
512	78.67	0.77	0.82	0.79	0.79
512	79.33	0.79	0.80	0.79	0.80
512	79.00	0.79	0.79	0.79	0.81
512	79.67	0.79	0.80	0.80	0.79
512	79.33	0.78	0.81	0.80	0.80
512	78.33	0.81	0.73	0.77	0.81
512	77.67	0.76	0.81	0.78	0.79

Table 8 shows the best result of CNN-GRU with w_dim = "64", Number of Kernel (n_kernel) was 128, Kernel Size (kernel_size) was 5, memory = "64" and Recurrent Dropout = "0.4". The result was accuracy 83.00%, precision 0.85, recall 0.80, f-measure 0.82, and ROC 0.83.

The best of CNN was accuracy 80.33%, precision 0.80, recall 0.81, f-measure 0.81, and ROC 0.80. The best of GRU was accuracy 82.33%, precision 0.81, recall 0.83, f-measure 0.82 and ROC 0.82, and the best of CNN-GRU was accuracy 83.00%, precision 0.85, recall 0.80, f-measure 0.82, and ROC 0.83. Therefore CNN-GRU is the best performance to classification sentiment with this dataset.

5 Conclusion

The purpose of this study is a classification of digital currency investors on twitter with three deep learning algorithms CNN, GRU and CNN-GRU. Dataset used as English and Thai textual from twitter 1000 sample and divided into 500 positive comments and 500 negative comments. After Preprocessed data then data split into a training set 700 samples and a testing set 300 samples. The classification results that CNN-GRU is high-performance accuracy 83.00%, precision 0.85, recall 0.80, f-measure 0.82 and ROC 0.83. In conclusion, a hybrid deep learning model is efficient more than a single deep learning model for sentiment classification. In the future, we would improve the performance of the model by using Transformer or Ensemble Architecture Model and explore pre-trained word embedding such as GloVe or FastText.

References

1. Tangwattanarat, M.N.: A study of the perception of Thai cryptocurrency investors towards digital currency market. Doctoral Dissertation, Thammasat University, Thailand (2017)
2. Xu, F., Zhang, X., Xin, Z., Yang, A.: Investigation on the Chinese text sentiment analysis based on convolutional neural networks in deep learning. Comput. Mater. Cont. **58**(3), 697–709 (2019)
3. Lee, J.S., Zuba, D., Pang, Y.: Sentiment analysis of Chinese product reviews using gated recurrent unit. In: 2019 IEEE Fifth International Conference on Big Data Computing Service and Applications (BigDataService), pp. 173–181. IEEE (2019)
4. Luo, L.-X.: Network text sentiment analysis method combining LDA text representation and GRU-CNN. Pers. Ubiquit. Comput. **23**(3–4), 405–412 (2018). https://doi.org/10.1007/s00779-018-1183-9
5. Makdee, S.: Big data analytic for tourism: a case study from Ubon Ratchathani Province, Thailand. J. Ind. Technol. Ubon Ratchathani Rajabhat University **10**(2), 139–152 (2019)
6. Luan, Y., Lin, S.: Research on text classification based on CNN and LSTM. In: 2019 IEEE International Conference on Artificial Intelligence and Computer Applications (ICAICA), pp. 352–355. IEEE (2019)
7. Shaohua, X., JingJing, L., Kun, L., Lu, W.: A parallel GRU recurrent network model and its application to multi-channel time-varying signal classification. IEEE Access **4**, 1–5 (2019)
8. Weng, L., Li, Q., Xuehai, D.: GRU based convolutional neural network with initialized filters for text classification. Aust. J. Intell. Inf. Process. Syst. **15**(2), 75–85 (2019)

9. Huang, C., Liu, G.: Sentiment analysis of network comments based on GCNN. In: Proceedings of the 2018 2nd International Conference on Computer Science and Artificial Intelligence, pp. 409–413. China (2018)
10. Gaye, B., Zhang, D., Wulamu, A.: A Tweet sentiment classification approach using a hybrid stacked ensemble technique. Information **12**(9), 1–19 (2021)
11. Wang, X., Jiang, W., Luo, Z.: Combination of convolutional and recurrent neural network for sentiment analysis of short texts. In: Proceedings of COLING 2016, the 26th International Conference on Computational Linguistics, Japan, pp. 2428–2437 (2016)
12. Poomka, P.: Deep learning techniques for sentiment analysis from product users. Degree of Master of Engineering in Telecommunication and Computer Engineering. Suranaree University of Technology, Thailand (2019)
13. Alotaibi, F.M., Asghar, M.Z., Ahmad, S.: A hybrid CNN-LSTM model for psychopathic class detection from tweeter users. Cogn. Comput. **13**, 709–723 (2021)
14. Zouzou, A., Azami El, I.: Text sentiment analysis with CNN & GRU model using GloVe. In: 2021 Fifth International Conference On Intelligent Computing in Data Sciences (ICDS), pp. 1–5. IEEE (2021)

Author Index

P. Meesad et al. (Eds.): IC2IT 2022, LNNS 453, pp. 195–196, 2022.
https://doi.org/10.1007/978-3-030-99948-3